Neural Almost Gate

How Neurons Shape Thought

Robert Hamill

A BT Book

Burning Thoughts Publications

I0116289

Burning Thoughts Publications
Columbia, MD
BTPublications@mentalconstruction.com
Hamill, Robert, 1947-
Neural Almost Gate : How Neurons Shape Thought
Includes bibliographical references and illustrations.
1. Thinking. 2. Creativity.
3. Learning. 4. Decision-Making.
ISBN 978-1-7332044-3-9
ISBN 978-1-7332044-2-2 (eBook)

Table of Contents

Dedication

A moment's insight is sometimes worth a life's experience.

- Oliver Wendell Holmes, Sr.

Preface

As a long-term Mensa member, I have always been fascinated by how we think and how individuals think differently. It frustrated me that explanations of creativity, intuition, and induction so often end with... the idea popped into the mind.

The reliance on words has never explained thoughts that are not logical, thoughts that leap beyond what logic provides. When I realized neural properties led to generalization and almost matches that are treated as exact matches, I found the angle to discuss non-logical brain processing and thoughts.

Over thirty years have passed since the germ of the theory occurred to me. In those years, I followed the progress of neuroscientists, cognitive scientists, and behavioral economists. For ten years, I've posted essays on aspects of the neural almost gate, but the presentation was piecemeal, not organized from alpha to omega. Three years ago, I started rearranging my ideas into ordered discussions.

Although I didn't intend to investigate the role emotions play in thinking, the neural almost gate developed a satisfying answer uniting genetics and experiences into emotions.

The final two chapters in *Neural Almost Gate* elaborate the relationship between preconscious, unconscious, and conscious thought and then their interlocking roles in behavioral choice.

On a final note, during the pandemic, I walked the paths of central Maryland. Sometimes when my glasses fogged up, I took them off. On one such day, I saw what I thought was a brown and green bush under a tree. Imagine my surprise when it stood up and walked away. The bush was, in reality, a woman in a brown sweater and a green skirt.

Neural almost gates fill in the blanks of our world, but perfection is not guaranteed.

Day 1. Taste of the World

Theorist: How does the brain work?

Tablemate: What? What did you say? Who are you? And why are you asking a complete stranger a question?

Theorist: Can I sit at your table?

Tablemate: Oh, I was so absorbed in my work that I didn't realize how filled the coffee shop had become. Sure. Let me move that album so you'll have room.

Theorist: Thank you. I'm visiting my cousin in the neighborhood.

Tablemate: Been here long? And why the strange question about the mind?

Theorist: They have put up with me for a week. Ten more days here, then I will submit my manuscript for publication.

Tablemate: You'll be right at home here. There are several writers that come in regularly. By the way, I'm a photographer.

Theorist: What jobs do you take?

Tablemate: Events that pay the bills. They fund my freedom to catch nature in action. That's my passion. Nothing makes me as happy as being able to capture an interesting event.

Theorist: I'm writing about interesting events, but from another angle. Do you believe the world exists as you perceive it?

Tablemate: So that's your interest in how does the mind work. Yes, I do. I know what I see and I know my photos capture it. See these shots from last month's Pavilion Art Festival. My clients pay me well to capture the feeling for them.

Theorist: I'm not doubting your skill, but can you recall times when your clients weren't satisfied by pictures you took for them? Perhaps they didn't like them.

Tablemate: True, but then their taste leaves much to be desired.

Theorist: Leaving tastes aside, that would mean that their minds didn't perceive the world as you did.

Tablemate: There's no disputing taste. Is that your point?

Theorist: No, it's not. Much knowledge is not absolutely true or false, but contingent on our personal perceptions, extended by theories and rely on moral judgements.[1] We need to separate raw facts from value-laden implications of those facts for a clear-eyed perspective on the information used to draw conclusions.

Tablemate: If you're trying to tell me that this coffee shop appears different for you than me, you're wasting my time as well as yours. They charge the same price to everyone who orders an iced latte and biscotti.

Theorist: I don't mean to suggest that physical reality is different for each of us, but we interpret the slice of reality we see differently. For instance, one of your clients may have warm memories of a family dinner at the pavilion, while another may recall an awkward first date there. As a result, the same picture will impact them differently.

Tablemate: Everyone knows that one's history casts a personal dimension on one's taste. Still, this doesn't make the world different.

Theorist: But it makes our internal world different from those of others.

Tablemate: Impossible. The world's the same for everyone. What's the big deal?

Theorist: We base our decisions on our perspective of reality, not the reality itself and that is where significant differences arise. I am writing about the neural almost gate, an investigation into how neural properties shape our thoughts.

Tablemate: Neurons, those little itty-bitty brain cells that allow us to think. They're common knowledge. Have you invented a brain calculus? That must be it. You sum up our neurons and out pop our thoughts. I'm excited to learn how millions of neurons add up to one little thought.

Theorist: You are widely off the mark. According to the current estimates, there are 82 billion neurons in an average brain, each having thousands of connections with others.

Tablemate: You're going to tell me what I think by examining my neurons?

Theorist: No, that's far beyond what I would dare to claim, but I do propose that neural characteristics cause regularities, which affect everyone's thinking because of our shared genetic ancestry.

Tablemate: Neural characteristics? This sounds like neural networks and computers. Are your ideas based on artificial neural networks used in computers, and not on the human brain?

Theorist: No, neural almost gates focus on the human brain, its neural structure, and the effects on our mental processing. Still, theoretical results sometimes prompted me to use artificial neural networks to investigate further.

Tablemate: Humph, but wait a minute. Sometimes you say 'brain' and sometimes 'mind.' Tell me what differentiates them as you see it.

Theorist: Sure. The brain is the physical assemblage of neurons, glia cells, neurotransmitters, neuromodulators, and the cerebral fluid inside your skull. The mind is the collection of experiences, beliefs, knowledge, and morals on which we base decisions and behaviors.

Tablemate: That's a catalog of items. What's the dividing line between brain and mind?

Theorist: The brain exists in the present, while our mind includes experiences which shaped it.

Tablemate: Since it's the brain holding those mental memories, your distinction is not absolute.

Theorist: True. Our mind occupies the same space as our physical brain. The distinction between brain and mind highlights particular functions, rather than physical separation. Brain refers to the physical structures and processes reacting to current reality, while mind adds to that memories and knowledge, expanding the current reality to past experiences and future possibilities.

Tablemate: Let's leave the realm of word-splitting. Tell me one idea about the neural almost gate, as you call it, that will surprise me.

Theorist: Sure. Feature matching guides our thinking as much as verbal analysis. Although many people insist that every idea must be logically proven, many thoughts arise from similarities and associations, not from logical deduction. This argument was first made by Aristotle (1952) in *De Poetica* where he argued:

The greatest thing by far is to be a master of metaphor. It is the one thing that cannot be learned from others; and it is also a sign of genius, since an apt metaphor implies an intuitive perception of the similarity in dissimilars. (p. 695)

Tablemate: Genius? Aren't we talking about you and me? We are ordinary people, not geniuses. So, tell me, what does a metaphor have to do with feature matching?

Theorist: Metaphors associate different things based on their similarity while ignoring any differences. For instance, a Jeopardy champ can be called a walking encyclopedia, as that person knows a multitude of facts. We use feature matching to make the metaphor, requiring only a close enough match, not an exact match.

Tablemate: I would say we count the number of features in each and the number of matches between them, then apply logic.

Theorist: But you'd be wrong. Feature matches arise from neural gates that treat patterns of close similarity as identical.

Tablemate: Different things can't be considered the same. You haven't even answered my earlier question. How do ordinary people fit into Aristotle's 'similarity in dissimilars'?

Theorist: I must tell you about the neural threshold and Hebb's Law[2] for you to understand why different inputs are treated the same, but that's too much for today. Let me answer the last question. Everyone, not just geniuses, uses the *intuitive perception of the similarity in dissimilars.* When we categorize, we notice that different categories share certain similarities. A simple example of this reduction occurs when we make jokes. We twist a commonplace notion in a new direction to get a laugh. That's near to a poetic metaphor. An aspiring genius uses a similarity, ignoring the dissimilarity in a new, surprising, and important domain. If the association results in success, then they are more than an aspiring genius.

Tablemate: Okay, maybe, but Aristotle says the similar in dissimilars is intuitive. That means unexplainable and beyond logic, doesn't it?

Theorist: Beyond logic, yes, but not necessarily unexplainable. There is a structure to the process of intuition which arises from feature matching or by associations supported by feature matching. This occurs through the neural threshold operation. Whenever the threshold is exceeded by the sum of input potentials, the same signal travels downstream to receiving neurons. As a result, closely similar patterns are treated as they were the same.

Tablemate: You want me to scrap logic and just accept any illogical association my brain comes up with. I cannot do that. You have distracted me from my paying project long enough. Leave me to my photos. No more neurons today.

Day 2. Neural Threshold

☕

Theorist: Let me tell you how properties of the neuron explain much of how we think.

Tablemate: Not that again. I only have fifteen minutes before I need to conduct my Zoom class. Astonish me. In one sentence, tell me how a hundred billion neurons give rise to my thoughts.

Theorist: I can't do that, but I can bring the answer a step closer. Let's start with Figure 1 depicting the structure of a typical neuron.

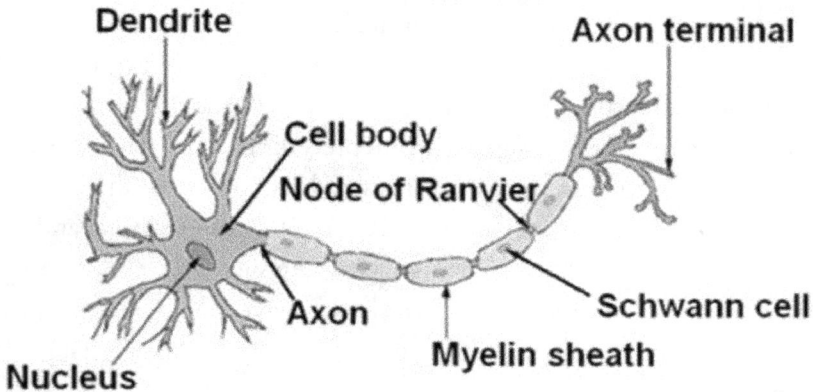

Structure of a Typical Neuron

Figure 1

THEORIST: THE NUCLEUS is in the soma, the neuron's body. The dendrites connected to the soma receive input signals from sensory cells or other neurons. All neurons work in the same basic way. If the total electric potential received by the dendrites exceeds the neural threshold, the neuron fires. It sends its signal down its axon to other neurons. The signal sent is the same, no matter which inputs contributed to triggering the neuron.

Tablemate: What about the other terms on the diagram? Schwann cells, nodes of Ranvier, and axon terminals?

Theorist: The myelinated Schwann cells wrap around the axon, forming small segments. Myelin sheaths insulate the electric potential of an axon signal, ensuring its rapid transmission. The gaps between the segments are nodes of Ranvier. At each node, the signal is regenerated, guaranteeing the accuracy of the signal delivered. In this manner, Schwann cells and nodes of Ranvier provide a link between distant neural layers.

Axon terminals are essential in this process, as they are engaged when the signal nears its destination. The signal splits, allowing each axon terminal to deliver the same signal to dendrites of the receiving neurons.

Tablemate: Are you saying that the axon signal arrives from another neuron via the dendrites on the left side of this image?

Theorist: That's right. Because only one neuron is shown in Figure 1, neuron-to-neuron connections are missing. Dendrites, on the left of the diagram, receive signals from preceding axon terminals. It's important to note that the axon terminal does not physically connect to the dendrite. There's a separation between the axon terminal and the receiving dendrite called the synaptic gap. The signal travels across the gap via a neurotransmitter, a biological messenger.

Tablemate: Just another step. When will you get to the end?[3]

Theorist: Not just any step. A crucial step. The signal is not transmitted across the synaptic gap with its full electric potential. When the transmission efficiency of the signal across the gap increases, that is a sign that learning has taken place.

Tablemate: Signals firing everywhere in the brain are expected, but how does that help explain our thoughts?

Theorist: Bear with me a little longer. It's a big deal that every combination of inputs that exceed the neural threshold causes the same signal to be sent to each of its receiving neurons, as it implies that a neuron does not differentiate sensations at an infinitely fine scale. If the inputs are nearly identical, the neuron's response is the same. It acts as if the inputs are identical. That different inputs give identical output astonishes me, as it results in dissimilar things being treated alike. If two disparate input sets enter a neuron and both cause it to signal, thereafter, these inputs are indistinguishable in the mind.

Tablemate: How dissimilar can the inputs be, yet be treated as the same?

Theorist: Excellent question. That's an answer I would love to see investigated, but to do so would require both experimental results and new mathematical techniques.[4] Still, there is no need to despair. Consider the Neural Threshold Gate shown in Figure 2. This example shows the huge number of inputs the neural body receives from its dendrites. The net sum of incoming potentials is the determinant of whether a neuron's axon will signal (or not) downstream to other neurons.

Neural Threshold Gate

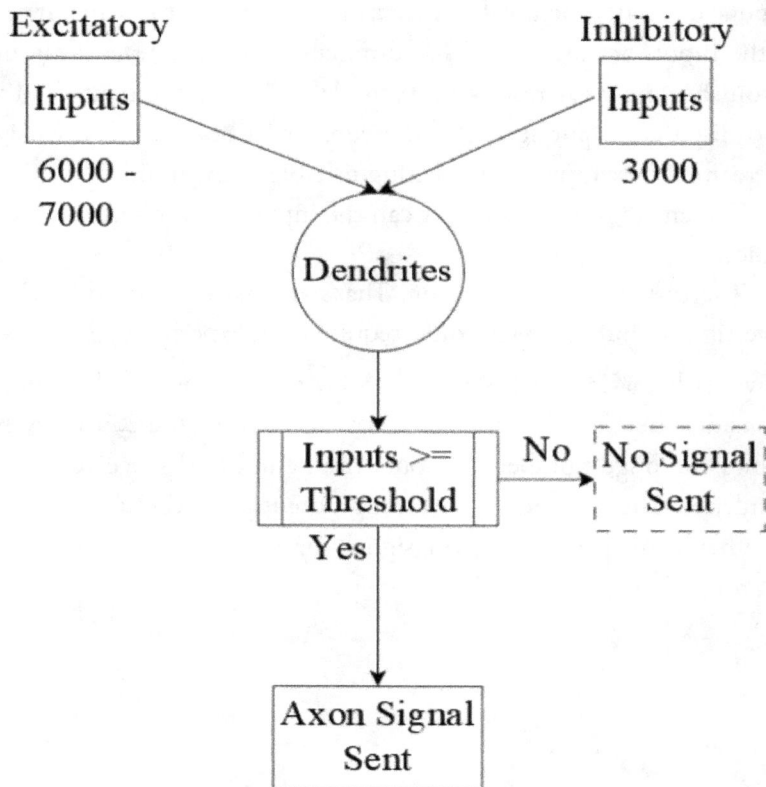

Excitatory   Inhibitory

Inputs   Inputs

6000 -   3000
7000

Dendrites

Inputs >=   No   No Signal
Threshold   Sent
Yes

Axon Signal
Sent

Figure 2

TABLEMATE: DO YOU INTEND to bury me in charts? Rectangles, circles, and lines prove nothing.

Theorist: Perhaps not, but they help organize the information visually. In this example, a typical brain neuron receives thousands of signals through dendrites. Its inputs are a mixture of excitatory and inhibitory signals. Both are essential. Excitatory inputs increase the likelihood that the neural threshold will be breached, while inhibitory signals decrease that likelihood. The inhibitory inputs sharpen the contrast between input sets, thereby promoting identification of input differences and facilitating learning. The diagram shows the neuron comparing the net sum of positive and negative electric potentials against its neural threshold. This summation depends on which dendrites have received a signal and how efficiently their neurotransmitters carried it across the synaptic gap. If the sum exceeds the neural threshold, the neuron fires. The signal is sent through its axon.

Tablemate: You're kidding me? There can be six thousand positive inputs and three thousand negative ones, with their sum determining the signal. That is not a very helpful simplification.

Theorist: Okay, but the output signal, when it occurs, is always the same. There is no sign of which inputs are associated with the signal. It just sends a single electric charge down the axon. The diagram shows that even if many thousands of inputs coalesce into a single output, the result will remain the same.

Tablemate: Different causes, the same outcome. It is going to take some time to digest that.

Theorist: Different causes produce the same outcome because of the neural gate, the almost gate. This surprising result has been confirmed experimentally.

Tablemate: No more! Things that aren't the same, yet are the same. This reminds me of my college class with philosophers debating the number of angels that could dance on the head of a pin.

Theorist: Have you heard of Bishop Berkeley's criticism of calculus? The Bishop criticized Newton's reasoning, saying it was based on ghosts of departed quantities.

Tablemate: That's funny.

Theorist: Yes, but it did not invalidate Newton's results.

16

Tablemate: My head is spinning. I must stop you here. I can't have metaphysical cobwebs while teaching on Zoom.

Day 3. Almost Gate

Tablemate: What are you typing this morning, my conversationalist?

Theorist: I'm trying to figure out how to explain the almost gate.

Tablemate: Isn't that the neural threshold?

Theorist: That's the basic form, but a layer of interconnected neurons also gives rise to almost gates. Neural threshold defines a single neuron's response. The brain's neurons, like those in the sensory receptor areas, are collected into layers that receive inputs from a related collection of sensory cells or neurons. Each neural layer processes its inputs across all the neurons in the layer, resulting in one winning neuron, which sends its signal to other neural layers. The winning neuron is determined by the highest magnitude registered across all neural thresholds in the neural layer's almost gate.

Tablemate: How is that an almost gate?

Theorist: With repeated exposure to similar inputs, the neural layer settles on one particular neuron in the layer as the 'winner' or marker neuron. This depends not only on the number of axon signals received but also on the synaptic weights of the interconnections. It's worthy of note that neurons in the winning neuron's vicinity respond to inputs similar in structure.

Tablemate: Words. That's just words. Unless you have proof, go somewhere else to spin your web.

Theorist: They are not just words. Maps of touch in the brain[5] were first discovered back in the 1930s by Wilder Penfield (Famous Scientists biography, 2022). In attempting to determine the way to treat patients with epilepsy surgically, Penfield found the neurons located across the top of the brain correspond to particular parts of the body. When he electrically stimulated a neuron area, the patient reported the sensation that would have been produced if the body part itself had been stimulated.

Tablemate: Okay. That's interesting, but maps are not evidence of almost gates. Perhaps neurons are hard-wired to those sensory inputs.

Theorist: We know that sensory input is not hard-wired to specific neurons. It's true that sensory cells relay their input to specific brain areas[6] or modules. However, in patients that had lost sight or a limb, neurologists have detected changes in the neural mapping by experimental methods similar to those Penfield used. They found that, in a blind person or an amputee, maps of new winning neurons were developed as a result of new experiences.

In this context, artificial neural networks are very helpful, since we can't experiment on a person's brain. In the 1970s, Teuvo Kohonen developed a neural network model in which all neurons in a layer connected to the same set of neural inputs as well as connected to all the other neurons in the receiving layer. That artificial structure mimics the cortical structure. Kohonen showed that repeated experiences not only led to the emergence of winning neurons, but also to the activation of neighboring neurons to respond to similar patterns. Admittedly, Kohonen networks are much simpler than those in the brain's hemispheres, but they confirm that feature mapping can take place with such structures. Look at the intriguing example illustrated in Figure 3A, E SOM Result. The Kohonen SOM (Self-Organizing Map, so called because no trainer or outside knowledge needs to be added for the input sets to be assigned to winning neurons. For simplicity, this example comprises of 35 input sensors, represented as a 5 × 7 matrix. Each of these 35 pixel locations connects to 100 neurons in the receiving layer, resulting in 3,500 connections. As each of the receiving layer's neurons connects to one another, that's an additional 4,950 connections. If nearly ten thousand connections for the artificial layer seems like a lot, consider that each cortical neuron has approximately 10 thousand connections with its neighbors.

After several hundred exposures to the dot matrix version of English alphabet, the computerized SOM neural network settled on a distinct winning neuron for each letter. In other words, it distinguished the letters without outside instruction.

In Figure 3A, the letter E fires its winning neuron in row 2, column 2.

E SOM Result

Sensory Data Standard E
5x7

Receiving Neural Layer
10x10

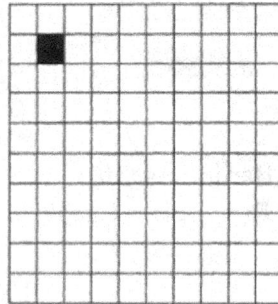

Each Input Pixel
Connects to Each Neuron in Layer
3,500 Connections

Each Receiving Neuron
Connects to All in Layer
4,950 Connections

Figure 3A

TABLEMATE: INTERESTING, the Self-Organizing Maps seem to distinguish letters, but I still don't see how this relates to the almost gate.

Theorist: The winning neuron does not match to one particular set of inputs, but to any set that produces sufficiently strong output to exceed the neural thresholds leading to its assignment.

Once the training was completed, the Kohonen SOM was presented with a novel input and produced the result Figure 3B. As you can see, the winning neuron produced by exposure to standard E still fired despite the visible difference between the shapes, as they shared enough similarities. The neural gate performed an almost match. Very evocative.

Off-Bar E SOM Result

Sensory Data Off-Bar E
5x7

Receiving Neural Layer
10x10

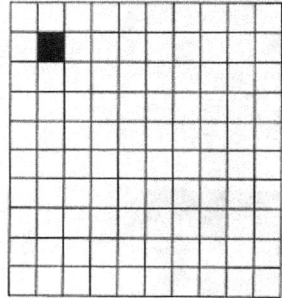

The Bar in E is Off
One Row Below Usual

Same Winning Neuron
As Standard E

Figure 3B

TABLEMATE: OKAY. INTERESTING, but I am not convinced yet. I need to get back to a restful work day at the coffee shop. Can you wrap it up?

Theorist: Yes, and I appreciate your patience while I struggle with explanations. Maybe you would care to hear that over 90% of cortical mass has the same neural layer structure as the sensory-motor region in which Penfield located the sensory map of our body. Therefore, it is reasonable to consider that even non-sensory layers of the cortex use maps to organize their information.

Tablemate: What do you mean? What are you trying to say?

Theorist: According to Manfred Spitzer author of *The Mind Within the Net* (1999, pp. 123-124), 99.9% of all neurons connect to other neurons, not to sensory or motor cells. Almost all processing in the brain involves abstracted—less detailed—information, rather than sensory information. Details are lost as neural almost gates process information. These abstractions are mapped too, but instead of relating to specific facts, they are maps of generalized information. The dogwood tree outside your kitchen window is a concrete image, but any dogwood tree is an abstracted image. Increasing abstraction of detail is a hallmark of further movement away from immediate sensory apprehension.

Tablemate: If I grant you what you haven't yet convinced me of, can you draw further conclusions?

Theorist: You were convinced, weren't you, that our neurons work with neural thresholds and that they abstract or remove excess detail when their electric potential barrier is exceeded?

Tablemate: It made sense. Yes.

Theorist: An almost gate is the collective action of all the neural thresholds in a neural layer after inputs are repeatedly experienced. It is collective because all neurons in the layer are connected through local axon branches to their neighbors. Experimental evidence indicates that they settle on a winning neuron for a set of repeated inputs. Likewise, nearby neurons will also win when presented with similar inputs. This shows that the entire layer acts as a map of the input range.

Tablemate: It is too much for me to accept now.

Theorist: Let me just say that mental maps are also known as categories and concepts.

Tablemate: Okay. But where did the first map or category come from?

Theorist: Hebb's Law.

Tablemate: Hebb's Law? I need more than a name.

Theorist: Sure. In 1949, Donald Hebb (Hebbian Learning, 2022) proposed a law to explain neural learning. He postulated that if the axon of neuron A is physically close to neuron B and neuron A repeatedly contributes to the firing of neuron B, the resulting synaptic changes will increase the signal transfer efficiency across the synapse.

Tablemate: That's Greek to me. Can't you explain it in English?

Theorist: The more frequently one neuron's firing contributes to another neuron's firing, the greater the strength of connection between the neurons. Remember the synaptic gap where the signal travels from the axon terminal to another neuron's dendrite? As I mentioned earlier, it is not the case that 100% of the signal crosses the gap. Under the conditions stipulated by Hebb's Law, the fraction of the signal crossing the synaptic gap increases. That increases the importance of the sending neuron's contribution to the potential firing of the receiving neuron.

Tablemate: Okay, but why is that relevant?

Theorist: An individual neuron only communicates two states, on or off (firing or at rest). A layer of neurons has a multitude of possible on–off states. Hebb's Law explains that experience will guide the layer to map the most frequent incoming patterns.

Tablemate: The barista is signaling that my latte is ready. My almost gate says it's time for me to apply myself to my client's project. Talk to you later—much later—scribbler.

Day 4. Handling of Information

Theorist: Would it surprise you to hear that our brain handles all inputs using the same neural processes?

Tablemate: No, and your assertion has no content that explains anything.

Theorist: Maybe things will become clearer if I add structural elements to the neural processes. Experimentalists have shown that our visual system proceeds through many steps—position, color, object identification, facial recognition, movement, and so on. While the visual system performs these tasks, the auditory, tactile, olfactory, and gustatory senses are operating in their specific neural layers of the cortex. The enhancement of sensory information involves many steps—some are sequential (such as those in each sensory mode) and others are conducted in parallel (those executed across the senses).

Once the current sensory environment has been received, enhanced, and categorized by the relevant cortical lobe, other significant steps take place as the perception travels toward consciousness.

Our brain's handling of data can be usefully divided into three stages, as shown in Figure 4A, Sensation to Category, follows a concrete example in which we apply our knowledge to physical stimuli.

Tablemate: Why do you insist on using drawings? Why not stick to verbal explanations?

Theorist: Because a picture is worth a thousand words in conveying the boundaries of ideas and because some people don't think primarily in words. I notice patterns and sequences, and then struggle to put them into words. You, I believe, think first in visual terms.

Tablemate: Perhaps. You can continue if you want.

Handling Information

Sensation to Category

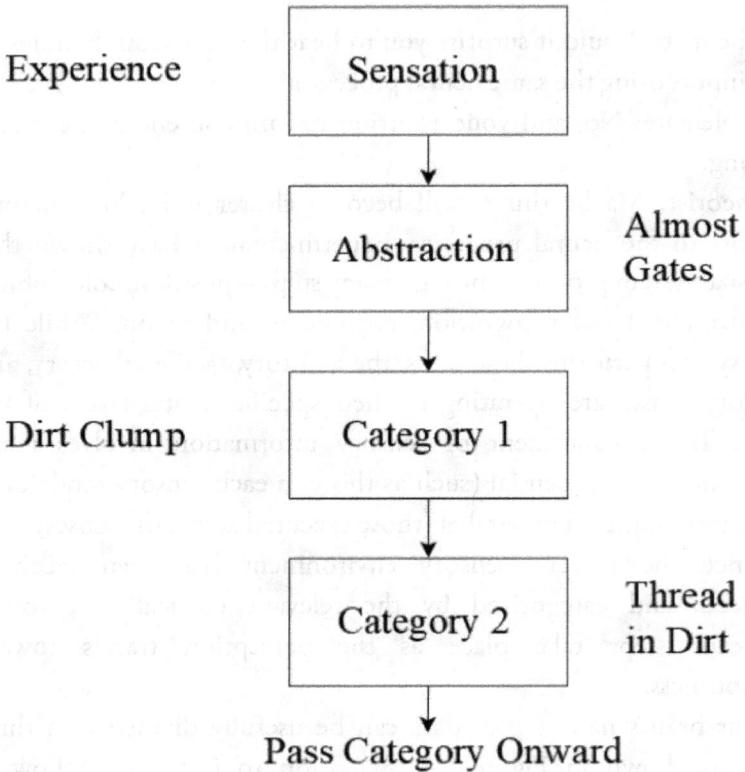

Figure 4A

THEORIST: SENSATIONS arrive in our cortex from sensory cells with binary on-off signals. An assemblage of binary states is received in a cortical layer, as we discussed the other day. Hebb (1949) demonstrated that, with repeated experience, wiring between neurons that fire together strengthens,[7] and the winning neuron's synaptic gaps efficiencies increase. Due to the more efficient transmission of electric potentials across the synaptic gap, winning neurons triggered by similar sensations will cluster about similar features.

In the receiving cortical layer, the winning neuron will be selected by an almost match, indicating that exact details are not delivered downstream. Sensory detail is abstracted, as details are ignored.

Once the first layer has processed the initial sensory input, the winning neuron passes its signal to the next cortical layer that summarizes specific layers into a wider range of the sensory field. This step results in a layer of integration and further abstraction.

The visual system has a wide set of cortical layers that have been experimentally identified, which are responsible for defining object outlines in our visual span, detecting color, position, motion, and so on. These processes occur automatically and are below our conscious awareness.

Information is further abstracted when the various senses are merged, as that involves passage through additional almost gates, as shown in a concrete example in Figure 4A. As indicated in the left column, if we see a thin, dense, dark material caked in dirt, we will likely ignore its precise shape and will perhaps disregard the variation of color and wetness across the material. Upon further abstraction, we may decide that we are looking at a thread caked in dirt.

Tablemate: I don't get it. Are you saying that everyone decides it's a thin thread when seeing a thin, dense, dark material caked in mud?

Theorist: No. The point is that people, without conscious thought, abstract from the details they perceive to what they consider important. That process depends on their genetic makeup and prior experiences. As shown in Figure 4B, the process continues as we add information based on learned knowledge.

Handling Information

Category to Concept

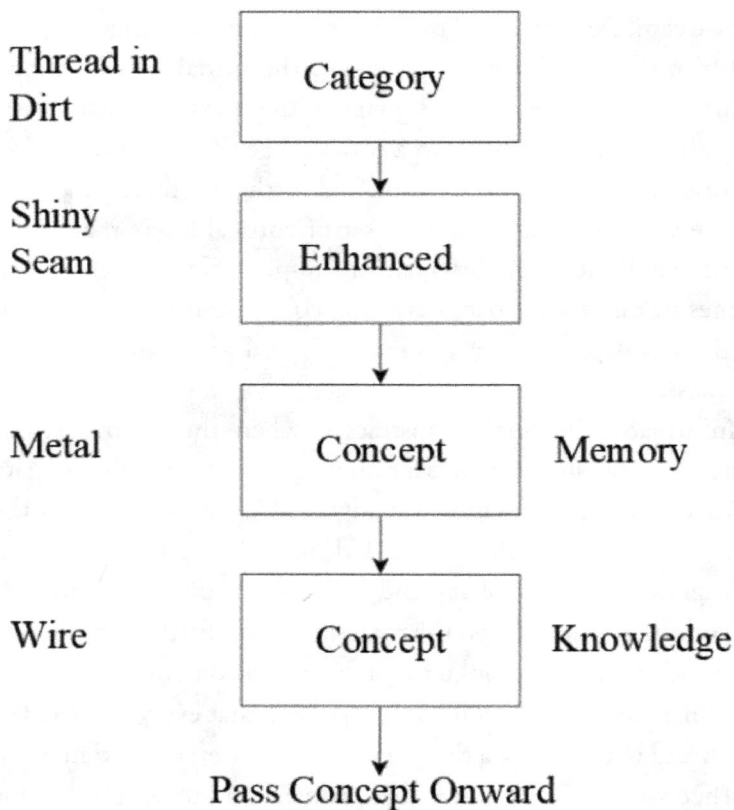

Thread in
Dirt

Category

Shiny
Seam

Enhanced

Metal

Concept

Memory

Wire

Concept

Knowledge

Pass Concept Onward

Figure 4B

THE DISTINCTION BETWEEN category and concept cannot be crisply drawn. The general idea is that category arises first and is driven primarily by one's physical experience. The concept is the result of organizing the category based on the acquired knowledge. When the category successfully surmounts the neural pathway of a concept, their properties are commingled.

The glitter from a dark thread caked in mud makes part of our neural process pass the same gates as a metal seam. As a result of this correlation, we would consider the thread a metal seam and, after further neural gates, we may class it as a metal wire. As the data proceeds through our cortical pathways towards the executive areas of the prefrontal lobes, the characteristics of metal wires adhere to the category transformed into a concept.

Tablemate: One thing troubles me about your explanation. In Figure 4A, sensations are narrowed down by removing details, while in Figure 4B, categories are expanding as properties are added. Yet you attribute both phenomena to the almost gates. Which is it? Do almost gates abstract or do they expand?

Theorist: You might hate this answer, but neural almost gates lessen incoming detail. They help us deal with the profusion of complexity in the real world, as a significant amount of detail is ignored, abstracted away. However, as the pulse of data flows through the brain, the magnitude of sensory detail decreases in relation to remembered knowledge. Although detail is lessened with each almost gate, the categories and concepts are connected to a progressively greater number of associated bits of knowledge, so that the absolute magnitude of information seems to increase.

Tablemate: How do particular properties of metal come into play?

Theorist: Let's take a look at Figure 4C, where concepts and associations are depicted.

Handling Information

Concepts and Associations

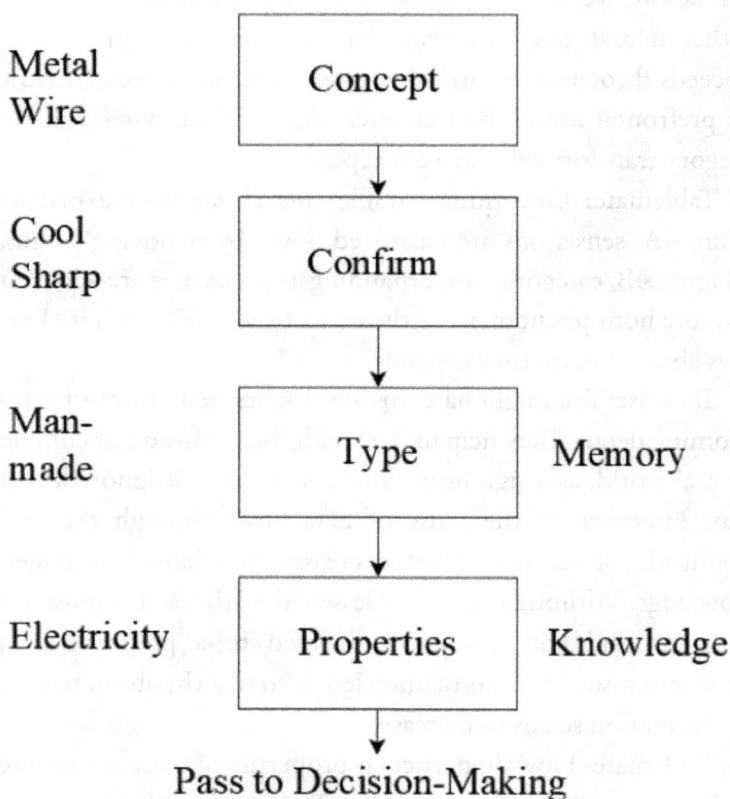

Figure 4C

THEORIST: ONCE A CONCEPT, like metallic wire, emerges after sensory input is processed, we would recall that it is a cool and sharp man-made object. In other words, the neural pathway the man-made metallic wire follows has been ploughed by one's acquired knowledge, as neural gates are primed by properties we associate with metallic wires, lowering the number of external stimuli for the property's winning neuron to be fired. For example, if we are aware that metallic wires conduct electric current, such a property will be more easily activated.

Tablemate: Let me get this straight. You are saying that, as a sensory input is processed, our personal experiences assign it to a category and then, using knowledge we have learned, we define it further as a concept. That concept also has properties, which we then associate with the initial sensory stimulus.

Theorist: You got it. That's it.

Tablemate: Wait a minute. I have to stop you here. I didn't hear a single mention of logic. How can you have thinking without logic?

Theorist: Most discourses on thinking focus on logic, so I have not focused on that in my investigations. Logical explanations of thinking overlook the inherent neural properties of abstraction, categorization, and association. The neural gate has guided thoughts down neural pathways long before the conscious brain evolved, and certainly long before the neocortex developed and gave logic to our thoughts.

Tablemate: But...

Theorist: A moment more, please. Almost gates give rise to inductive thought, the ability to generalize from a set of facts to a general rule encompassing all considered facts. Logic alone cannot explain induction, intuition, and their cousin, creativity.

Tablemate: What! How's that?

Theorist: The brain is full of concepts of varying levels of detail. They are abstracted categories, weighted by personal values and augmented with knowledge. When these concepts, enriched categories, reach the prefrontal lobes they can be shuffled in working memory. The executive sections of the prefrontal lobe have great flexibility in retrieving categories of various abstraction. Only at the highest level of thought does logic come into play.

Tablemate: Let me get this straight. The ultimate step is the only place where logic plays a role?

Theorist: That's right, but I readily admit induction can lead you astray. Logic absolutely has an important place in thought. Inductive decisions may look sensible, but it can cause us to overlook some important factors that were lost in the abstraction steps. Despite these sometimes false conclusions, associations have one great advantage over deductive logic, as they generate new ideas, wrong though they sometimes may be.

Properly applied, deduction always leads to valid conclusions. Yet, in many important situations, all the facts needed to complete a logical argument are not available. To compensate for this loss, we substitute inductive associations for facts and go ahead with our logical calculations.

Tablemate: Wait. You switched from induction to associations with too much ease. How are they related?

Theorist: Sense data is categorized, like the pink dogwood tree in the backyard to dogwood, then to trees in general.

Tablemate: So what? And why does it lead to other trees?

Theorist: Because when one specific tree is stripped to its essential treeness, it is mapped to the same winning neuron as other trees. At that point, it becomes the same for further reasoning.

Tablemate: I hesitate to ask, as you might go on forever, but I need an example of why you think induction or association replaces logic in our thinking.

Theorist: I am not saying that deduction is never used. Often, we use association or induction to supply a missing fact or proposition to which we then apply deduction. Daniel Kahneman (2011) in *Thinking Fast and Slow* labels this tendency "the availability heuristic." It is very useful, as it helps us take an easy-to-access fact and substitute it for the exact fact. That's an association. This reasoning can be applied to the stock market, as there is no accepted theory that forecasts stock prices.

To invest in the stock market, a person makes assumptions. A technical analyst compares recent price movements with those documented in earlier periods. If the analyst notices similarities and the earlier result was lucrative, the analyst performs an induction and asserts that the beneficial relationship will continue. That's their technical basis for recommending stocks. Price movements of the past become facts on which we apply logic.

Tablemate: If all the signals are positive, it doesn't take an analyst to decide to buy the stock.

Theorist: That may be, but that logic is still only based on associations between past and current experiences, which may not continue. Abstracting data as if minor differences don't matter is not a logical process. It results from a winning neuron chosen in a neural layer through an almost gate.

Tablemate: I've got another question. How is induction related to the almost gate?

Theorist: When categories lose detail and gain aspects of remembered experiences, they share features with past events. When these associated memories travel through other almost gates, such differences progressively diminish until the associated concepts become the same. The result is an induction from specific experiences to a generalized association.

Tablemate: That's too much for me. The words are too nebulous.

Theorist: I'm sorry that I'm not clear. Let me leave you with this thought. In science, often thought to be purely logical, hypothesis selection results from induction and insight. That inductive step guides the experimental parameters as well as the analytic methods, which depend on logic. Logic and induction by almost gates have advanced science for hundreds of years. Does that help assuage your worry that I'm slighting logic?

Day 5. Concept Elevator

Tablemate: Good morning, Mister Loquacious. How's the writing progressing?

Theorist: It's coming along. Can I add an iced latte for you to my morning cold brew order?

Tablemate: No, thank you. I'm on my third green tea, but have a seat anyway. I need a break from examining these photographs of eagles at Conowingo Dam. My students will pepper me with questions about composition, lighting, contrast, and myriad of other topics. Their questions will be innumerable, so your ramblings are a welcome change.

I did think of something I wanted to ask you. Why are almost gates so important? Everybody knows that if I come in here with a new haircut, I'm still the same person. It's just a superficial change.

Theorist: Good. I'm glad to hear you admit that. Almost gates are important because they exist throughout the cortex, not just for sensory input but also for recalling memories and retrieving knowledge.

Facial recognition requires more than a single neuron's almost gate. It requires an interconnected neural layer, each receiving the same visual information. The neural module responsible for facial recognition is in the fusiform gyrus, a cortical area below the occipital and temporal lobes. Within it, a winning neuron is assigned for each face it recognizes. If a person is unknown, that winning neuron may be shared by many other unfamiliar faces. What is significant here is that the winning node in the recognition module can be selected by closely related inputs. There is not just one perfect match, but once the winning neuron is selected, our thoughts go ahead as if the almost match was exact. I'm sure you've mistaken one person for another.

Tablemate: Yes, of course. Everyone knows these things happen.

Theorist: We use experience and knowledge to develop our internal worldview, which leads to the concept elevator.

Tablemate: You may never convince me of your ideas, but I do enjoy a good metaphor. What is the concept elevator?

Theorist: It refers to the changes that occur as external phenomena are processed from sensory reception to conscious awareness. The sensory information is augmented by memories of prior experiences and knowledge, as well as by our needs, goals, and fears. By the time the sensation arrives in the prefrontal lobe, this augmented situation contends with our current internal worldview, which guides our decision-making. This is exemplified in Figure 5A where the concept elevator takes us from sensation to category.

Concept Elevator

Sensation to Category

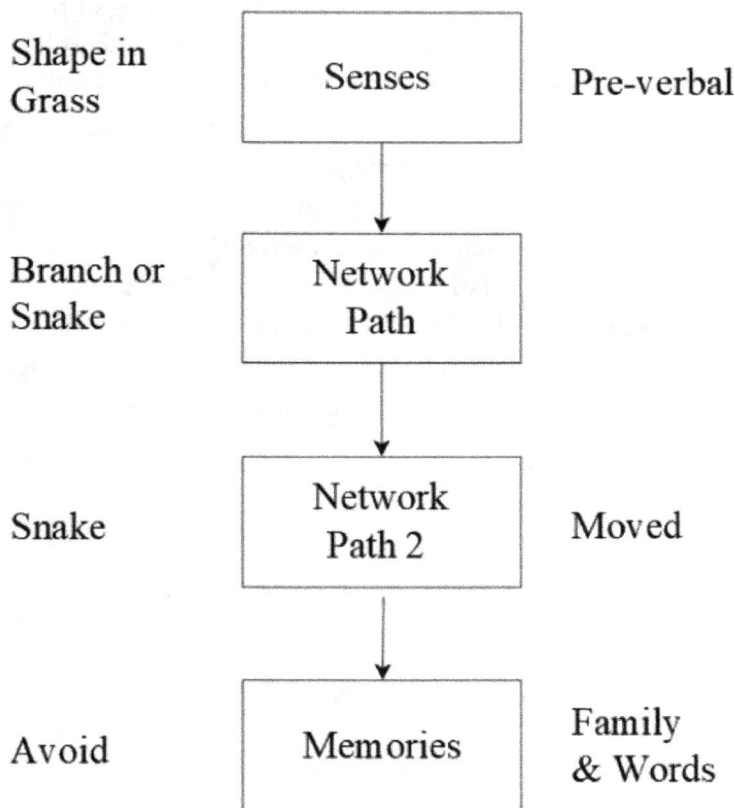

Shape in
Grass

Senses

Pre-verbal

Branch or
Snake

Network
Path

Snake

Network
Path 2

Moved

Avoid

Memories

Family
& Words

Figure 5A

TABLEMATE: THAT DRAWING looks like yesterday's Sensation to Category figure.

Theorist: Not quite. The emphasis shifts to the forces shaping of categories as information flows through the cortex.

The senses box stands for numerous almost gates, which form our pre-verbal observations. This processing was developed long before humans had invented language. Infants use it prior to learning speech. The next box shows that the pre-verbal observations are then organized by family interests and verbal categories. The functioning of the boxes in the right column will be developed further when we discuss learning and even further in the discussion of the brain's maturation. The implication is that you learned some extremely basic patterns (shape, smell, touch) before acquiring language.

In the left column, I used a slightly different example than before to demonstrate the changes in the concept elevator's progress. Our initial category is a long, thin, dark shape in the grass. As a child, we learn from our parents and siblings that, if it moves, it could be a snake.

You may wish to look at Figure 5B where the concept elevator moves from category to situation to understand my arguments better.

Concept Elevator
Category to Situation

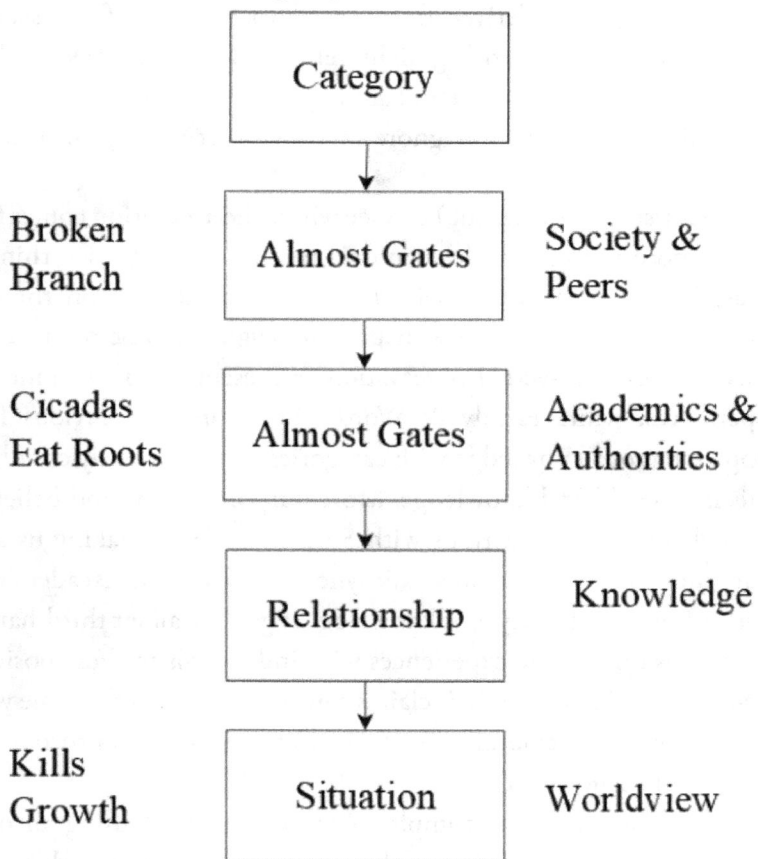

	Category	
Broken Branch	Almost Gates	Society & Peers
Cicadas Eat Roots	Almost Gates	Academics & Authorities
	Relationship	Knowledge
Kills Growth	Situation	Worldview

Figure 5B

THE CATEGORY PASSES through an array of almost gates. The arrow from Society & Peers points inward to the almost gates, as these learned categories have shaped the efficiencies of synaptic gaps in the almost gates that make certain neural pathways more likely to be surmounted by the inputs.

Tablemate: I'm confused by the boxes. What are the implications?

Theorist: The boxes represent activity at some stops of the concept elevator. They're stops at which the raw input is changed by the synaptic weightings derived from prior experiences. Of course, these figures are simplified to highlight relevant features. They mask the actual complexity of the cortical activity.

Tablemate: I get it. You ignore anything you can't explain and plow ahead.

Theorist: Not quite, but I can see where the accusation comes from. My purpose is to show the neurons' role in our cognitive thinking. Thus, I ignore side avenues if they have little impact on the main thrust—the external world is shaped and augmented before it reaches consciousness. Pre-verbal observations represent our first, immediate experiences, while Family & Words shown in the previous figure supplement our knowledge with categories learned in early childhood. This is second-hand knowledge. More purported facts and beliefs are learned through interactions with Society & Peers, making us aware that the Family view was idiosyncratic. Through Academics & Authorities, another layer of knowledge is gained, albeit third-hand, as most of us don't share experiences with individuals in these positions. Hence, the reliability of their claims can't be assessed in the same way as the information we obtain directly or learn second-hand through our family and friends.

I have provided an example in the left column, using an object recognized as a branch. We could directly verify that it didn't move, and a friend told us there was a storm last night while we slept. Based on this information, we concluded that what we were observing was a branch blown off last night. The dark object in the grass has changed from a category, an object observed, into a concept, an object organized by knowledge. This concept travels further through our learned knowledge, including that delivered by Academics & Authorities. In a recent article about the cicadas, the authorities informed us that cicadas feast on tree roots, killing new growth. This information stimulated by the thin, dark shape in the grass becomes part of the situation arriving in the executive lobes.

Tablemate: I'm not sure I like this thinking of yours. You claim that each of us thinks the world differs from actuality and each one of us perceives it with a different slant. Thus, according to this argument, everyone acts in their own subjective world. There is no absolute right or wrong.

Theorist: Yes, each person acts on a subjective worldview, but societal interactions with the outside world have led us to rules and norms that shape our sense of morality and impact on our decisions.

Tablemate: But...

Theorist: Morality is not a direct consequence of neural gates. Morals are a cultural determination that we learn through social and academic interactions, but the neural cascade has implications bearing on the subjectivity of one's internal worldview. It is another consequence of the almost gate.

Tablemate: Neural cascade? Another metaphor?

Theorist: Yes. As the concept elevator updates the stream of information heading for conscious awareness, each update passes numerous almost gates. For the moment, forget about the content change as the information moves across the brain. Think of each step passing almost gates, triggered by less than a perfect match. At each gate, detail is lost. A neural cascade of fidelity loss to the original sensory perception.

Tablemate: That reminds me of my earlier question. How much detail is lost at an almost gate?

Theorist: Have you heard of the 100-Step Rule (Schmidt, 2000)?

Tablemate: Is that like the F-stop on my camera?

Theorist: I don't know about cameras, but the 100-Step Rule is a rule of thumb in the cognitive realm. Experimentalists have established that we need a minimum of 0.5 seconds to react to an external stimulus.

Since individual neurons react in roughly five milliseconds, Feldman and others have proposed that about 100 neural firings or steps must occur between sensory input and bodily reaction. Half of those steps, 50, are inward along the concept elevator, towards one's internal worldview. The remaining 50 steps are the neural firings on the path to performing our reaction.

Tablemate: Suppose I grant 50 steps or stops on the concept elevator to consciousness. What's the point?

Theorist: At each of those 50 stops, a slight loss of fidelity occurs, resulting in the next stop starting with a less precise sensory picture. Let's just say that the loss at each almost gate is 1% of its input detail. For simplicity, we will ignore the alterations imposed on the situation because of our experiences, needs, goals, and fears. After 50 steps, we've lost about half the sensory detail.

Tablemate: Wait. Lost half of the image! I don't think so. And where did you get the 1% loss at each almost gate?

Theorist: To address the second question first, I must admit that I made up the 1%. It's undeniable that different patterns can cause the same effect, as previously discussed in relation to facial recognition. How much would you say a person with a new hairdo matches the same person with their old look? I just took a guess at 1% to allow me to make a quantitative argument. I'd much prefer to have experimental evidence on the size of input exactness required to select the winning neuron, but that's not available.

Tablemate: Am I to understand that speculation and imagination are the base of this argument? It's easy to contradict your assertion. The coffee shop around us is not missing half its detail.

Theorist: I cannot argue about exact percentages, but you will admit that, unless you focus on the specifics, you cannot answer how many tables there are, where they are placed, which ones are occupied, who is sitting at them, and so on. You only have a general picture of the coffee shop. As a photographer, you will likely see more details than I would, as you have trained yourself to notice visual details I ignore, but you would still miss out on many.

Tablemate: There may be a little truth to that.

Theorist: Granted, the fidelity loss at each almost gate is not precisely known. Of course it would be productive to know the exactness of the required match, but the consequences of my 99% match and 1% difference guess remain illustrative. The neural cascade results in an inevitable loss of fidelity because of neural processing through successive almost gates. Since the height of the almost gate is driven by individual genetics, creative people with lower thresholds, will see relationships between many things, lose more fidelity than those with higher almost gates like dot-the-i-cross-the-t thinkers, will make fewer errors than free-thinkers.

Tablemate: I don't buy that. As a photographer, I am creative, yet I want to capture what is there, not what I imagine might be there.

Theorist: Let me recast that. Our mind doesn't just respond to the present, but it prepares for future opportunities.

Tablemate: I have noticed something. Your explanations suffer from the same effect you are trying to explain. For instance, if two input streams surmount the same almost gate, it doesn't matter that they differ in insignificant details. So, you can't explain why one feature is more important than another. There is no logical hierarchy to the explanations as there are no logical relationships to learn.

As a result, you find it hard to explain the most important concepts, because all features are equally important if they contribute to surmounting the almost gate. They meet the same feature matching criteria. You can't put sentences and thoughts together in a logical hierarchy that organizes your explanation. I'm surprised I could make sense of any of your points. Sense, perhaps a little, but not logical sense!

Theorist: You're right. It's hard to organize these ideas. Still, I am encouraged by the fact that you've agreed with a few points I have made so far.

Tablemate: I'll leave you to wallow alone in lost fidelity. My photographic reality is calling. This morning's dialog will fade into the mists of a neural cascade. You have helped confirm that I prefer objective truth to subjective escapism.

Day 6. Learning

Theorist: Good morning. I'm working on a tricky question. Would you mind if I talked it through with you?

Tablemate: Isn't that my fate these days? Sure, it'll be lost in the neural cascade soon enough. I have time and you're already talking. What's the question that's bothering you?

Theorist: How do we learn? How do we learn the patterns, categories, concepts, and ideas that our internal worldview is furnished with?

Tablemate: You're kidding, aren't you? Learning. That's easy. Get an excellent teacher, study, and do your homework. You'll learn.

Theorist: That's the way to academic learning, but that's only one learning mode that involves explicit knowledge acquisition which is heavily influenced by verbal descriptions and conscious consideration. But I want to talk about implicit learning[8] or the ability to learn skills and how things interrelate. This implicit learning developed, over the eons, as the brain dealt with environmental challenges.

Tablemate: Okay, okay. You mean something else. Does that photo of bird footprints have something to do with it?

Figure 6. Squiggles or Letters.

THEORIST: I USED SQUIGGLES or Letters for the Figure 6 caption, as the image reminds me of learning to read. When I was very young, my mother used to read the Sunday comics to me. *Peanuts* was a special favorite. After many, many Sundays, I realized that the squiggles of Woodstock's footprints and the squiggles in balloons above the characters' head played different roles. The balloon captions held words that described what was said or thought, although the footprints looked similar to letters making up those words, but weren't words.

Tablemate: A nice memory, but what's the point?

Theorist: The learning process is the point. I learned not because mom explained the alphabet, but because her finger pointed to the balloon squiggles as she read. That's learning by induction—the repetition of an event until I recognized a relationship. With time and repeated experience, I came to recognize more and more squiggles as letters and gradually progressed to relating them to short words she read.

Tablemate: Let me guess. That's implicit learning because you weren't shown standard letter shapes like teachers write on the blackboard, but rather learned to associate their shapes to the relevant sounds. You generalized that they went together.

Theorist: Precisely, and I did that without conscious awareness. It's only in retrospect that I recall it happened without effort.

In neural terms, we group each shape and sound into particular categories. These groupings result from a series of repeatedly fired neural layer[9] almost gates. They are not logical conclusions based on rigid rules, but an inductive recognition of feature matching. Do you remember the extra difficulty of distinguishing 'b' and 'd'? More experience helped you master the distinction.

Tablemate: You mean that noticing that an 'A' has two slanted lines meeting at a peak with a cross-bar is an induction, not a logical conclusion of its definition?

Theorist: Until we started formal education, induction was our primary learning mode, although once we started speaking, verbal communication allowed us to learn from others. Induction remains important even for an adult. Consider all the computer fonts that exist. You recognize the English letters no matter how fancy or plain they may be in a font because you recognize the letters by their salient features, not by exact memorization.

Tablemate: I won't argue about medieval fonts that have so many curlicues that I am not sure which letters are which.

Theorist: I'm glad to hear that. That shows you learned the alphabet by induction. You recognize an 'E' by its features,[10] not its precise shape. And when not enough features match, you can't decide.

Tablemate: Still, excellent teachers are essential for learning.

Theorist: You continue to limit learning to explicit knowledge, lessons we learn by verbal instruction. Let's discuss more fundamental learning, like physical skills and social knowledge. Walking, talking, and getting along in groups. How do we learn that?

Tablemate: Those skills are easy to gain through trial and error. First, you learn to stand, then put one foot in front of the other. You may fall, but soon gain confidence and walk independently.

Theorist: You are being flippant, but I'm serious. These are implicit skills that we first learn by attempting and observing others, not by instruction. Later, we may form a verbal description of those skills, but that's after the fact, not the learning mode.

How do our leg muscles know when to tighten and when to extend? What about the rules of our society? When do we realize what's right or wrong, allowed or forbidden, eccentricity or lunacy? How do we first learn words and how to pronounce them? How do those skills and morals get into our brains?

Tablemate: Again, through trial and error, as well as by observation.

Theorist: Okay, but what internal changes must occur to support these skills? What changes when we learn? Perhaps you've seen Eric Kandel on the PBS Nova program. He discusses how he verified Hebb's Law experimentally by measuring physical changes occurring at the synapse of a sea slug, Aplysia californica, during learning.

Tablemate: I saw him at the start of a program, but a friend called and I missed it.

Theorist: That's a shame. Dividing learning into before, during, and final phases is useful. Let's start with Before Learning shown in Figure 6A.

Learning

Before

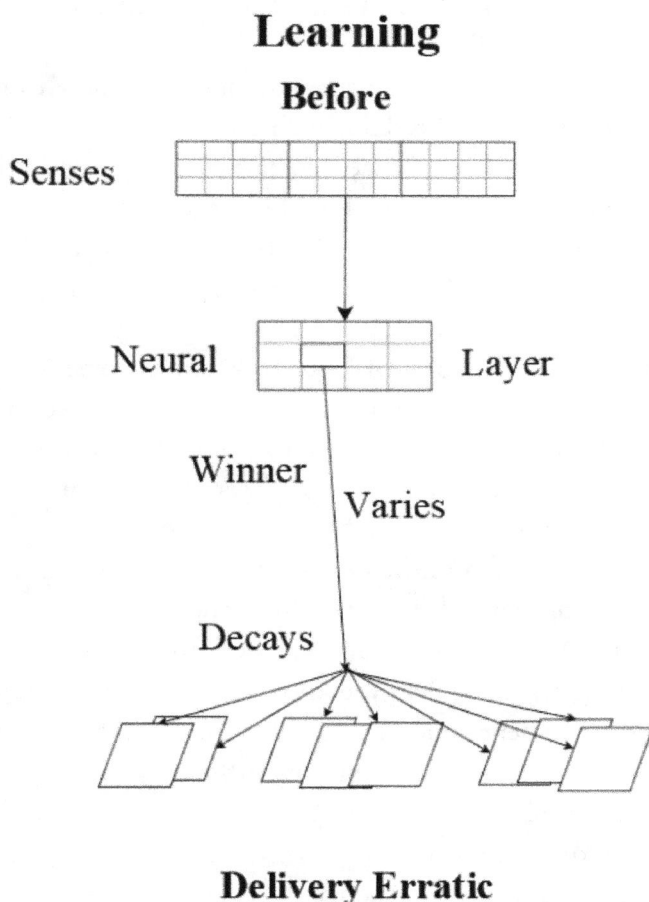

Senses

Neural ⬚ Layer

Winner

Varies

Decays

Delivery Erratic

Figure 6A

DID YOU KNOW THAT IF we don't learn certain skills at specific ages we will never learn them properly?

Tablemate: Yes, I know about the critical learning period. But everyone I know can walk and talk, and many can even chew gum at the same time. The critical learning period had no impact on them. As for learning, many people that take my classes are well into their sixties and seventies and are still learning.

Theorist: The critical learning period does not apply to explicit learning, but rather specific skills, such as language processing and social skills. Our age, circumstance, and particular experience when we learn implicit skills play a strong role in the categories and concepts that thereafter are used to organize explicit learning and make decisions. This will become clearer as we consider how the brain processes information prior to the critical period.

As shown in the top part of Figure 6A, inputs come into neural layers from sensory cells or from other neural layers, but we will focus on sensory data here. Consider visual information from the optic nerve being received in the occipital lobe's V1 layer. As the winning neuron for the same input set is not fixed before birth, a different neuron may win each time. The axon of the winning neuron sends a signal, but it often decays before being delivered to distant neurons.

Tablemate: No one expects a fetus to see in the womb.

Theorist: That's funny. Did you know that the vision system starts its critical development period as soon as the infant is born, but other skills remain in the "before" stage for months and years, like walking, language, and social skills?

The start of learning is depicted in Figure 6B using language development as an example. About a year after birth, an increase in Gamma-Aminobutyric Acid (GABA) occurs. GABA is an inhibitory neurotransmitter that suppresses neural activity distant from the winning neuron. This allows learning to begin. Through repeated experiences and in accordance with Hebb's Law, the local neural layer develops a preference for specific winning neurons for the experiences it receives.

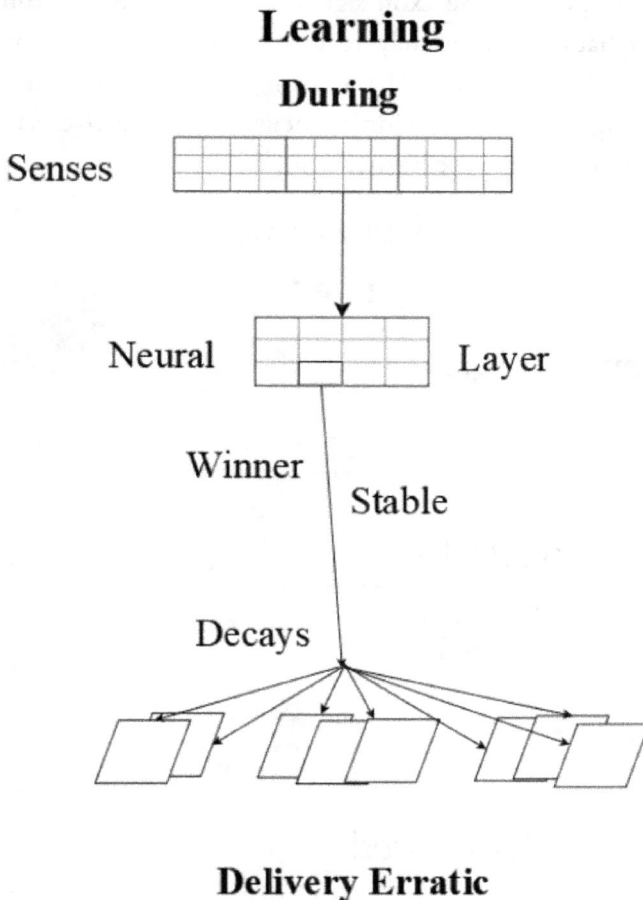

Figure 6B

FOR LANGUAGE ACQUISITION, this learning process continues for several years, with the child recognizing more and more words, but this categorization of auditory input is not used by other neural layers.

Tablemate: Why not?

Theorist: Because the axon signal of the winning neuron is not delivered reliably or accurately to other neural layers in the brain. However, with myelination of the axon, the critical period ends, as shown in Figure 6C. The winning neuron signal arrives at distant neural layers, delivering the learned result.

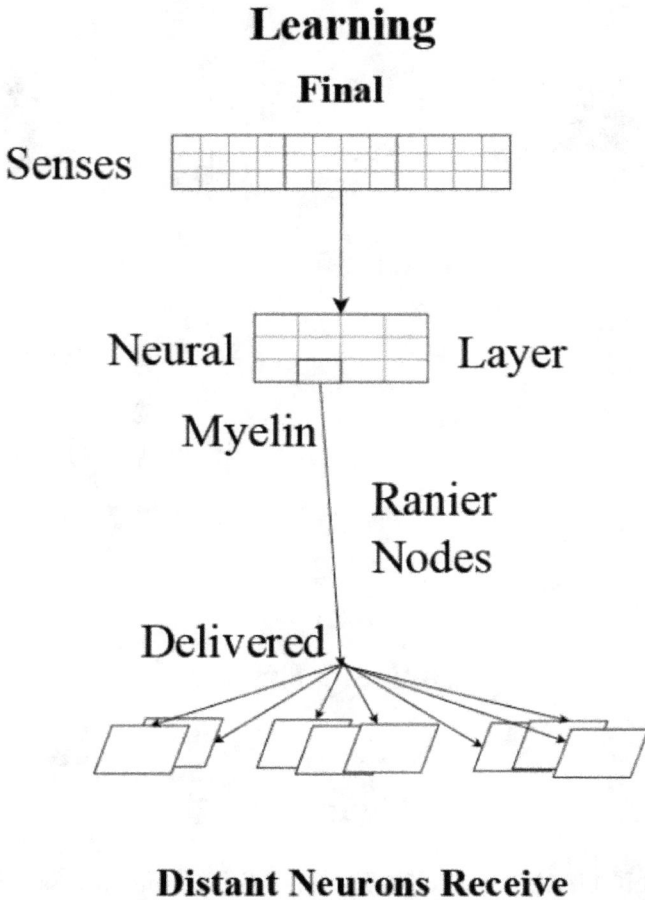

Figure 6C

TABLEMATE: WHAT ENSURES that the transmission is now delivered?

Theorist: Two developments are required for this outcome. First, as GABA levels decline so does the recognition of novel input. Fresh input is categorized to existing winning neurons, since synaptic efficiency cannot easily change with diminished GABA.

Another change occurs along the axon, which is insulated through myelination, allowing the signal to travel faster and further without degradation. The Nodes of Ranvier are a related change along the axon. They regenerate the signal, like a repeater in a telephone circuit. Through laboratory measurements, scientists have established that these two changes allow the axon signal to travel ten times faster than previously and deliver the output signal with greater reliability. When the implicit learning phase ends, the categories one uses are set and delivered to other regions of the brain.

Tablemate: If I understood you correctly, we learn language and maybe a skill or two in this implicit way. So what?

Theorist: Let me return to the critical learning period, as you have acknowledged that it exists. I would add that, once the critical learning period has lapsed for a cortical area, the set of categories is fixed and delivered to other neural modules. The learned categories provide the groundwork for downstream modules during their critical development periods. As a result, any quirky categories may cause unusual learned categories in downstream modules. Examples can range from color blindness affecting our visual organization to a cleft palate affecting our social behavior because of others' reactions to it.

Tablemate: That reminds of the song. There was a man who found a crooked sixpence and lived in a crooked house. That type of quirky thing? What makes a quirk? How do we learn something that is a quirk?

Theorist: Yes, that is what I meant, and he bought crooked nails to fix his crooked roof. If the category floor tilts one way, the second floor may tilt in compensation. Identification of quirks is difficult on the sensory level, because those quirks often leave the person physically impaired to react. Let's consider a social skill instead. Assume that, when you were a child, your parents explained to you that every social problem was caused by negative immigration policies. Once you can understand what that means, that becomes a feature of your youthful categorization of social events. As you age, have more experiences, and create new experiences as categories in your worldview, those new categories will be built upon the original categories of injustice suffered by immigrants. The original categories don't disappear because you have learned new higher-level and more abstract categories. The earlier categories remain, with patches, extensions, and exceptions, as a part of your worldview.

Tablemate: You have an entire pattern to explain every case, don't you?

Theorist: I wish that were true, but sadly I don't. There are many types of categories and their distinction from concepts is beyond my skills to articulate.

Tablemate: Wait a second. Category appears to be your catch-all term. I'm not always sure what you mean by category and how that differs from concepts.

Theorist: Categories, as a term, describe patterns at various stages of abstraction. Sometimes we use a concrete category like a gray 2011 Toyota Camry. At a more abstract level, the category may be "car." Still higher, and joined with our emotions, the concept may be fear that a new driver may be involved in a car accident. All these categories originate from the same physical sensation, but they reflect a different level of abstraction.

Tablemate: Why didn't you say that from the start?

Theorist: I tried to, but wasn't very successful, was I? After the crucial period, the relationship between categories and input features is gelled. The role of a local neural layer shifts from learning new categories to providing fixed categories of its information to the downstream neural layers. New inputs received after the critical period has ended are not learned, but are rather forced into existing categories. As a result of this restriction, considerable distortion between the raw input and the winning neuron's axon output to other layers may occur.

Tablemate: Is this right? The critical learning period is like a window of opportunity. If people don't learn something during the critical period, they will never learn it. That has interesting implications for why people may be blind to new ideas.

Theorist: I'm pleased you see that. In *The Other Brain: From Dementia to Schizophrenia*, Douglas Fields (2009) provides this useful summary of myelination which marks the end of critical learning periods: "myelination proceeds in a slow wave from the back of the cerebral cortex to the front as we reach adulthood" (p. 282). Our critical learning periods proceed from visual, auditory, and bodily control skills, through social skills, and culminate in conscious thought and executive decision-making.

Tablemate: Wait. Executive decision-making is last. Next, you'll say we should let young mass murderers go free because they weren't responsible for their crimes, as their minds weren't yet complete.

Theorist: Whoa! I believe incarceration should match the action, not the intent or reasoning. But that's a legal issue and is not what I'm discussing here. My aim is to explain how learning in one stage affects the next stage.

Tablemate: One second. Thinking back, I just realized you're saying that we need to learn to see.

Theorist: Yes, one has to learn to interpret what the eyes present to the occipital lobe. The visual system has a critical period that spans from birth to two years old. Vision is enhanced in the rearmost cortical lobe, by the back collar. Research on children born with retinal cataracts shows that indeed there is a critical learning period for developing vision. The results show that the earlier cataracts are removed, the more normally the child's vision will develop. However, if the removal occurs after the age of two, despite receiving retinal input, the children never learn to identify what they see. They are functionally blind.

Tablemate: Humph. I suppose the same is true for hearing and language.

Theorist: Yes, and it also holds for social skills. Howard Klawans M.D. (2000) in *Strange Behavior: Tales of Evolutionary Neurology* and other books, has published insightful essays on feral children who came into social settings during critical learning periods for language (age 1–8) and socialization (age 3–12). One child heard human speech for the first time at the age of six, yet she learned language easily when given the opportunity. Another child was twelve when he first heard human speech and sadly he never learned to speak or understand language.

Tablemate: Wait. I have another question. What about criminals? They violate societal rules by using violence. Are you saying they missed the critical learning period for socialization?

Theorist: No, most criminals have not missed socialization, but rather learned skills that violate our social norms. The critical learning period explains broad threads of how we learn, think, and act, but given the great disparities among individual experiences, we must acknowledge that general trends of human thought can be twisted by unique and odd environments.

Tablemate: You can't say anything specific about anyone, can you?

Theorist: Unfortunately that's true, but I can make arguments that apply to most people, and can discuss general mechanisms that organize their unique experiences.

Tablemate: Give me an example?

Theorist: The words you learn before the age of eight will remain the fundamental vocabulary of your thoughts forever more. In grade school, your vocabulary will expand. It will expand further in high school, at college, and later at work. Your word store continues to grow into adulthood. Yet, when you ruminate, perhaps mulling over the events of the day, considering impacts of social skills learned as a child, you will likely revert to your primitive first words. They are the floor of your verbal universe; however, I do not mean that higher executive thoughts are restricted to this primitive reality. The tension we often feel between our desires and our verbal conclusions are a sign that earlier categorizations persist through our lifetime.

Tablemate: Is there any proof of this?

Theorist: The staged nature of critical learning shows that early categories lay the foundation upon which later learning is based; however, that's a general argument not specific proof. Fortunately and unfortunately, the recent boom in neural science has provided more avenues for research than cognitive scientists can pursue. Language acquisition studies have focused on the prediction of future academic success, not on how one's initial vocabulary affects later language usage.

Consider this question. No need to answer aloud. Why do you think some people find it hard to understand gender and sexual categories that are not binary—man or woman? Our earliest categories are altered by later learning, not erased. People explain new concepts in terms of those they already know.

Tablemate: I have another concern about learning. Isn't drug addiction a result of learning? Same with alcoholism, gambling, and overeating?

Theorist: Yes, but those are examples of unfortunate learning. I am focusing on learning with positive outcomes.

Tablemate: Okay, but doesn't your theory have to explain the negative outcomes as well to be considered a successful theory?

Theorist: That is a good point and one I have not considered. Thank you for pointing it out and, of course, for letting me try out my thoughts with you.

Tablemate: At last, we agree on something. Now leave me alone. I must prepare for this afternoon's Zoom conference.

Day 7. Brain Building

Theorist: Good morning. Today, I'm working on an exciting idea—the steps from no brain to the first brain and then to emotions and consciousness.

Tablemate: Big words, but exciting? I don't think so. As the brain got bigger, we got smarter.

Theorist: You're too glib. Different life forms were breathing and eating long before consciousness emerged.

Tablemate: One second. Barista, bring me an iced latte and chocolate biscotti. He's paying. Now, tell me scribbler, what difference does it make that long-forgotten creatures knew how to eat?

Theorist: Because they are our history. Structures in our brain, like the brainstem, are not so dissimilar from those of the early invertebrates. We control bodily functions like heartbeat, respiration, and digestion in the oldest portions of our brain. Only after those processes do we use our newer brain modules to choose our actions.

Tablemate: You may think like a Neanderthal, but I don't. You are describing what high school biology called autonomous functions. Even their name shows they are not under conscious control.

Theorist: You're right. That's a better way to put it. The autonomic nervous system is controlled by parts of the brain that work without conscious awareness. As for Neanderthals, they are practically our twins compared to the first organisms.

Tablemate: Go on while my ancient brain feeds itself.

Theorist: The rise of the human mind has three significant steps, as shown in Figure 7A.

Rise of Human Brain

Reflex to Homeostasis

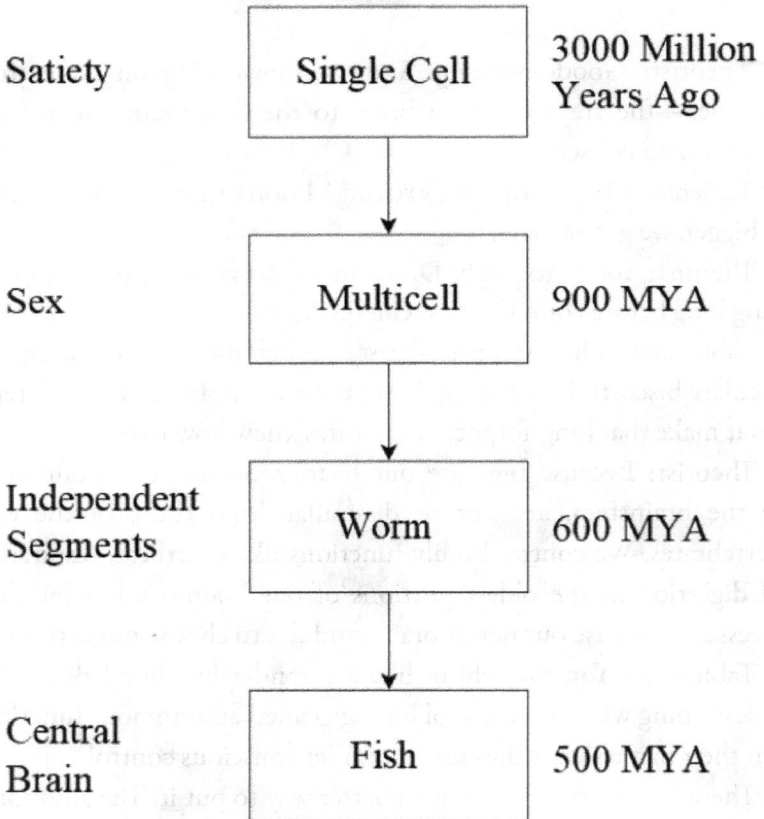

Satiety	Single Cell	3000 Million Years Ago
Sex	Multicell	900 MYA
Independent Segments	Worm	600 MYA
Central Brain	Fish	500 MYA

Figure 7A

THEORIST: AT THE TOP are single-cell microorganisms that had existed at the beginning of life on Earth. They arose over three billion years ago and were capable of simple behaviors. When one encountered an object, it automatically attempted to digest it. Its second behavior manifested when the single-cell microorganism exceeded its optimal size, which would prompt cell division, giving rise to its duplicate. As this was a unicellular organism, its one cell performed all the behaviors of the cell. A complex biochemical reaction directs the organism to eat. Let's call that satiety. Neurons, special-purpose communication cells, have not yet made an appearance.

The unicellular microorganism reigned until multi-cell microorganisms appeared some 900 million years ago, which had differentiated cell structures, including neurons. These neurons existed between the microorganism's sensory receptors and its cells that allowed movement, but no centralized control (as in a brain) existed.

Many multicellular microorganisms produced sexually. Think of that. Nearly a billion years ago, sex was a factor in the actions of many living organisms on Earth. It's a deeply embedded imperative.

Tablemate: It's not surprising that sex is important.

Theorist: The third life form depicted in Figure 7A arose after several hundred million years of evolution and responses to environmental changes. It is at this point that the first invertebrates, worms, appeared.

Tablemate: Excuse me. Nothing in-between?

Theorist: No, I didn't mean that. The fossil records show that there were many intermediate organisms, but I am only mentioning those most revealing of brain development.

More than a half billion years ago, worms made their appearance. They were physically much larger than prior multicellular microorganisms. These worms had nerve nets, not centralized in a brain, but dispersed across the worm's body. This arrangement allowed different sections of the body to experience external stimuli differently and react independently. A limited communication between nearby nerve cells allowed a rhythmic pulse to propagate along the worm's length.

Tablemate: Like the wave in a sports stadium.

Theorist: Yes. Not long after the worms in the span of life's development, but still 500 million years ago, a significant step forward occurred. Fish appeared. Sensory and motor cells remained near the fish's exterior, but for the first time, many neurons gathered into a centralized brain. The other neurons were connected to the brain by a spinal cord. The spinal cord also delivered to the brain the status on the fish's internal organs, such as lungs, blood contents, and digestion.

The fish's brain consisted of the pons, medulla, and cerebellum. These structures are called the old brain in humans. The pons and medulla exist deep below our cortical mass, deeper than the limbic system, while the cerebellum exists as a separate mass behind the other brain mass and supported the fish's movement. The pons and medulla controlled basic functions like breathing, heart pumping, blood glucose and metal content, eye movement, and so forth. Homeostasis, maintaining internal conditions for the organism's best performance, developed in fish hundreds of millions of years ago.

Ready for Figure 7B, where I presented emotions and consciousness?

Rise of Human Brain

Emotions and Consciousness

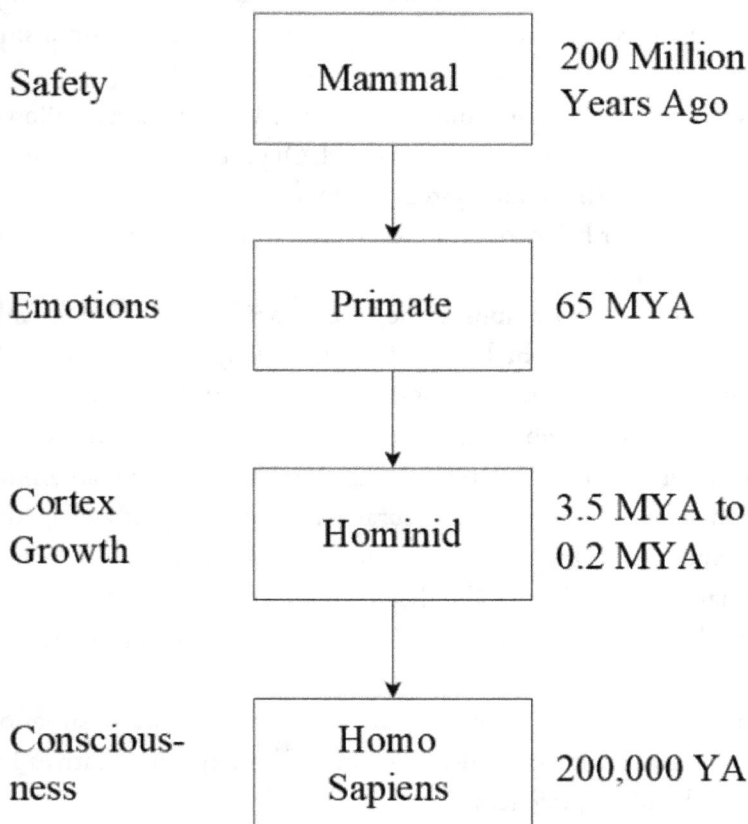

Safety	Mammal	200 Million Years Ago
Emotions	Primate	65 MYA
Cortex Growth	Hominid	3.5 MYA to 0.2 MYA
Conscious-ness	Homo Sapiens	200,000 YA

Figure 7B

TABLEMATE: MAMMALS, primates, and people. Finally, things that have meaning to me. That abbreviation, MYA, in the right means million years ago?

Theorist: Yes. To simplify the figures, I skipped the amphibians which emerged between worms and mammals, 340 million years ago.

Atop the fish brain, the amphibian's had a forebrain. The increase in neural mass allowed combining separate homeostatic demands in a single response. Instead of an invariable response to the latest information, the amphibian brain received updates from multiple bodily organs continually. O course, it could only react with a single behavior at any particular moment.

With mammals, additional brain structures appeared, allowing greater flexibility in reaction to the world. Of particular importance for our discussion is the limbic system's hippocampus and amygdala. They are critical to our behavior and the manner in which the human brain works and thinks.

Until the hippocampus came along, animals didn't remember events, for they only had immediate knowledge of situations. The hippocampus gave mammals long-term memory, and thus the ability to save the information about their environment, their reactions, and the actual results, into the enlarged cortex. For the first time, an animal's reaction would flow through neural pathways efficiently shaped by prior experiences. In psychological terms, mammals could weigh the potential success of a possible behavior by considering the results it previously yielded. Memory enabled safety to become a factor in decision-making.

The amygdala is the origin of our emotional reactions to situations. Emotions reflect our preconscious view of how our imperatives have been met by life experiences.

In neural terms, our previous experiences and their consequences travel through neural pathways, adjusting synaptic efficiencies. After many years of experience, the almost gates settle into a stable pattern of most likely activation when a similar situation is encountered.

The so-called 3S Imperatives

—Satiety, Sex, and Safety—form the bedrock of mammal behavior. They are not choices, but drives that our behavioral choices aim to satisfy. It is very rare that all 3S Imperatives are satisfied completely by the same behavioral choice. Choice typically involves a trade-off between the three drives. Although mammals do not have speech, we label the significant trade-offs as emotions. For instance, lust implies that sex is the sole imperative being satisfied. Love satisfies most of sex and safety, while gluttony is an over-satisfaction of satiety with a price on mammal's physical well-being. The amygdala processes emotion in the limbic system and delivers the emotion to the small cap of cortical neurons which act as intermediaries between the limbic system and the final motor cells necessary to exercise the behavioral choice.

Tablemate: One second. I need another chai. Do you want anything?

Theorist: No, thank you. I'm on a roll.

The emergence of primates 65 million years ago ushered in significant growth to the simple cortical covering of the mammalian brain. The new cortical neurons were interconnected with the existing brain neurons rather than to sensory and motor cells. Due to these neural layers, sense data could be associated into a more holistic picture of the external world.

As shown in Figure 7B, the first hominids emerged about three and a half million years ago. Tremendous growth in the cortical mass occurred as the hominids evolved toward homo sapiens. The tripling of neurons in the cortical lobes led to sensory enhancement.

The frontal lobes expanded. Although not involved in enhancing sensory input, they supported alternative consideration, that is, executive decision-making. The significant increase in neurons allowed for many potential behaviors. The primate was still guided by its needs, goals, and fears, but it could imagine the consequences of an immediate reaction as well as devise multiple behavioral paths that could lead to satisfying its 3S Imperatives.

Consciousness, as I use it, is not a catch-all term for what otherwise can't be explained, but is defined as the ability to distinguish between current actual reality and future imagined reality.[11] With humans, consciousness had been achieved.

Tablemate: Are you implying that consciousness is merely the ability to tell now from the future?

Theorist: That is the core distinction required to have consciousness and is necessary for this discussion of the neural mind's work. The qualia, the sensations and experience of consciousness, will have to be addressed by others. I'm ignoring that avenue to stay focused on my core concern: how do neural properties affect decision-making differently than verbal and logical deductions?

Tablemate: Okay. I can ignore that for now, but back to the amygdala and emotions. Why are there so many emotions and not just three?

Theorist: The 3S Imperatives are often satisfied only in part. It's possible to satisfy hunger by picking berries from a bush shadowed at the edge of a rocky hillside, while ignoring sex and accepting an uncertain tinge of danger. If that combination of satisfaction occurs often enough, the cortical brain applies a name that differs from satisfying hunger while with a sexual partner and no hint of danger. There can be as many emotional states as experiences of partially satisfied primitive imperatives.

Tablemate: I see. You've broken down emotions into simpler components. That might have some limited worth, but I have to meet someone at the library in a half-an-hour. Let's wrap this up. What would you consider mandatory for me to understand your theory of the rise of the mind?

Theorist: I can finish before you need to leave, but bear with me a bit longer. Consider this. Only within the past 200,000 years have humans evolved enough cortical capacity to think consciously. Throughout the ages, preconscious actions reigned and mostly satisfied the 3S Imperatives.

The older parts of the brain have fewer layers and fewer connections than within the later evolved cortical lobes. That simplicity results in preconscious thoughts being more granular than conscious thoughts. Subconscious thoughts, which arise in the limbic system that can sometimes be interpreted by words, are also supported by fewer layers, resulting in more black-and-white thinking. That is, fewer extenuating circumstances are allowed for in our mental considerations.[12]

Tablemate: I think that's good. A fact is true or false, nothing else. That's the law of the excluded middle.

Theorist: Yes, but the law of the excluded middle is limited to actual facts, actual occurrences. In most thoughts and discussions, we rely on opinions, beliefs, and knowledge.

Tablemate: Why do you have to make everything so complicated?

Theorist: Because the world is complicated. Consider this simple question.

Was the war in Afghanistan a wise decision? That's not a factual question, but an evaluative one. The answer depends on your values as much as the facts about the fighting. Your values are often yours alone, not shared by other citizens or by citizens in other countries. My point is that many aspects of a thought process other than pure facts go into answering a question. Many statements connecting facts are evaluations, beliefs, or theories and predictions about the relationships between facts. Those statements are outside the True–False division of the law of the excluded middle.

A logical approach to the question is "Was the war in Afghanistan a sound decision[13]?" Answering this question implies one's ability to reach a valid logical conclusion based on given premises and assumptions. However, a sound conclusion also demands that the premises are true. Only if they are true the conclusion can be considered knowledge. If, in fact, an assumption is false, the logical conclusion remains valid, but applying it to the real world is a mistake.

Tablemate: I don't see how a logical conclusion could be wrong.

Theorist: It's easy to be fooled. Consider this assumption—most immigrants commit crimes in our country. If you agree with that statement, you are led by logical argument to support restrictions on immigration. Yet, when one researches the subject and discovers facts that don't support the base assumption, any logical conclusion, though valid, would be deemed unsound, false, and unsupportable.

Tablemate: I reject immigration restrictions logically.

Theorist: I'm glad to hear it. But I assume you arrived at that position because you considered the facts underlying the assumption. Many adults don't question the assumption, and can be easily swayed by valid arguments based on unsupported facts.

Tablemate: A fact is a fact is a fact. How does your argument relate to grainy subconscious-level thoughts?

Theorist: Despite re-aiming Gertrude Stein's aphorism, if you use a fact in a theory, it's the fact that remains indisputable, not the theory. Regarding granular thoughts, the limbic system, source of emotional determinations, has less mental bandwidth to weigh extenuating circumstances. Emotions are more granular than conscious thoughts.

As we age, the emotional beliefs we formed in earlier stages of life are blunter than our conscious mental abilities, and yet those emotional beliefs continue to guide our decision-making.

Tablemate: Did you notice my eyes glaze at your mention of stages and emotional beliefs? Your words need to connect to the reality of daily life.

Theorist: Can I impose on your time a few more minutes? I want to show you Figure 7C depicting development of adult thought, as it may answer some of your questions.

Rise of Human Mind
Development of Adult Thought

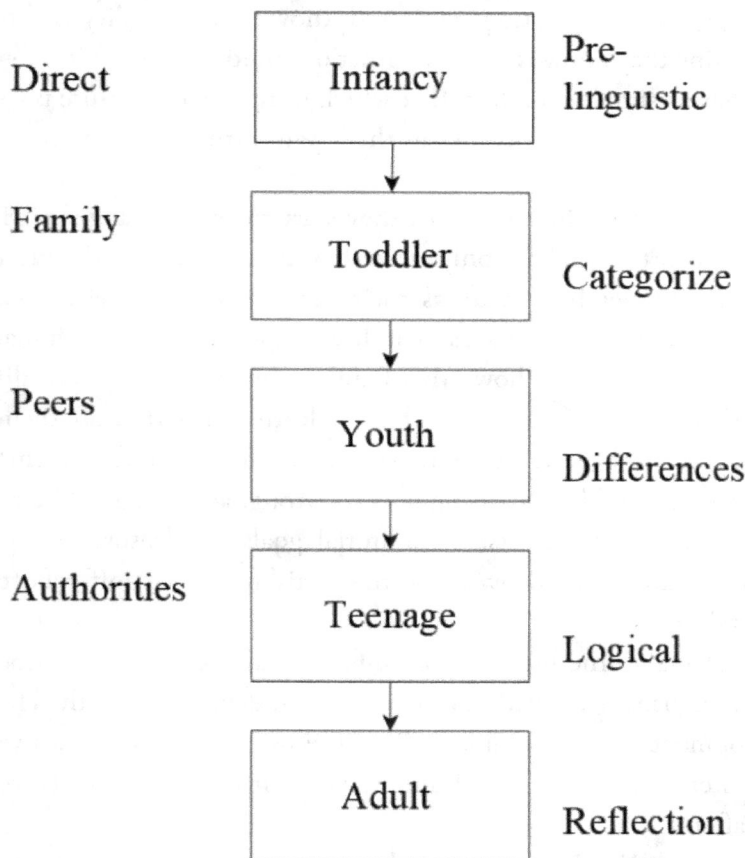

Direct	Infancy	Pre-linguistic
Family	Toddler	Categorize
Peers	Youth	Differences
Authorities	Teenage	Logical
	Adult	Reflection

Figure 7C

TABLEMATE: FINISH YOUR damn explanation. If I get a ticket on the way to the library, you have to pay it. Quick now.

Theorist: Adapting Kohlberg's stages of moral development[14] (Encyclopedia Britannica, 2022) and merging it with the relative immediacy of knowledge reveals an intriguing layering of knowledge and morality in the formation of adult mind.

As shown in the top box, given that infants are born without language, their interactions with and knowledge of reality is direct. Everything they know is a fact. It either occurred to them or it did not. They have a simple morality. That which brings them pleasure is to be demanded and that which brings them pain is to be avoided or cried about.

As infants age into toddlers, they start to acquire language. This allows them to learn not only what they experience directly but also what family members relate as their experiences. This second-hand knowledge is not as reliable as their direct experiences, but with shared experiences, they learn how other family members' descriptions differ from their own impressions. As children learn words, they use them to categorize their experiences, both for decision-making and for memory. In this stage, children learn right from wrong, according to the rules imposed by their parents. Their initial goals of pleasure and pain avoidance are overlaid by a reward–punishment tradeoff reflecting parental wishes.

Tablemate: The line between infancy and toddlerhood is not so firm. Language is learned at different times, depending on the child's development and environment. Plus, the use of language takes years to master in primary school and, if you consider new words, years thereafter.

Theorist: Yes, I ignored that blurring effect and emphasized the differences to make the points clearer. For instance, when youngsters meet neighborhood friends and start attending school, their peers expose them to different family norms. They realize that their family rules are not universal. Youngsters also hear tales of experiences that they can't verify or assess. With these new experiences, their certitude about facts and speech is shaken. At this point, a new decision-making metric emerges—group approval or ostracism.

Tablemate: Childhood and the next box, Schooling, are hardly distinct. They usually occur simultaneously.

Theorist: True, but they introduce knowledge of different immediacy, evaluated by separate criteria. Teachers present information that we can't test and evaluate, though we can often judge the validity of stories our family and peers tell us. That's why I've listed them separately. In addition, schooling continues after childhood has ended.

Tablemate: I hope you understand I am not nit-picking, but merely pointing out how much ambiguity you are asking me to accept when weighing your argument.

Theorist: I take your point and wish I could cleave the stages more crisply, but our life proceeds along multiple paths at once. When youngsters join their friends at the playground, they continue to directly experience the swings, slides, and the climbing apparatus while categorizing them with words as well as verbally sharing experiences. The earlier methods of learning haven't disappeared because the stage had ended. They continue to be used and are augmented by the latest stage's developing prowess.

As children grow into teenagers and young adults, they learn in school, at work, and through social interactions. They learn that societal rules and norms govern most behaviors. Teachers and the media expand their scope of knowledge, and they are able to distinguish theoretical, academic, and belief-based information. However, the ability to judge third-hand experience as true or false is diminished as students have minimal shared experiences to assess the reliability of authority figures that are presenting different facts and theories. By early adulthood, logic can be used to assess morality. Is the behavior logical and consistent with societal mores?

When they reach maturity, people often start re-examining their societal beliefs and actions. If the results of societal actions seem undesirable, the person's moral compass often adjusts to resist or condemn social norms.

Tablemate: Didn't you say your theory doesn't address morality?

Theorist: Not exactly. The theory doesn't prescribe what morality should be, but it does describe choices with moral aspects.

Tablemate: Do you mean I'm not mature, because I don't protest all day, every day, against every injustice in society?

Theorist: I didn't mean it that strongly. I mean, a mature person doesn't automatically accept societal guidelines as inviolate. Some may be wrong or misapplied. They can be challenged.

Tablemate: That's better, but your explanation of stages was long-winded with the major points were unclear.

Theorist: That's why I start with a figure, so that by referring to its elements in my explanations, I may make them easier to understand.

As we age, the role of consciousness moves to the fore. Thereafter, our knowledge grows through continued exposure to second- or third-hand sources. However, unlike knowledge gained by our personal experiences or related to us by people we know and whose trustworthiness we have ascertained, such third-hand information is difficult to verify. To use this new knowledge, we need to develop a new skill—how to decide which authorities—teachers, media, scientists, religious figures, etc.—to trust.

From birth up to the stage in which we strive to become part of social cliques, learning is mostly unconscious, occurring in the more ancient regions of our brain, like the limbic system and amygdala.

Tablemate: Enough. A final question, then no more. What significance does the development staging have for how we think?

Theorist: The guidelines for the morality of behavior you learned in the earlier stages do not disappear as consciousness grows. The earlier attitudes remain and become the assumptions we draw upon to categorize experiences. These assumptions relate not just to the events that have occurred, but also to how we believe our actions are related to different consequences. It is only with much mental energy and determination that adults scrutinize earlier presumptions for their continued veracity and usefulness. Often we don't make the effort.

Tablemate: Before we part, I have another question for you hotshot. What does my decision to leave now tell you? That my meeting at the library is a moral imperative?

Day 8. External to Internal

Theorist: It's so hectic in this coffee shop. One day it's quiet, but the next day it's chaos. There's a different crowd all the time. How do I keep my inner calm in an ever-changing world?

Tablemate: I don't see it that way at all. The world is always the same, yet my mind jumps from one thought to another, as my focus never seems to last long.

Theorist: Well, you can write a book from that perspective if you ever feel like it, but I believe that the external world changes more often than one's internal world. As we live and experience life, our internal world becomes more defined and fixed. That's why, as we age, we despair that the world is becoming nonsensical. I ascribe this position to the internal worldview being more stable and rigid, and thus more difficult to change.

Tablemate: Your statement reminds me of *Mindset: The New Psychology of Success*, in which Carol Dweck (2007) explains that everyone has either a growth mindset or a fixed mindset. People with the growth mindset believe they can master a task with enough effort, while those with the fixed mindset believe that, if they fail the first time, it's because the task lies beyond their ability.

Theorist: I like Dweck's ideas. They are very insightful, but her analysis is not at the neural level. She argues that early childhood experiences help shape a person's mindset, which affects their choices throughout life. I agree with that perspective. It bolsters the argument that the stage at which one gains specific knowledge shapes the person thereafter.

Still, as I promised that I would use images to convey my ideas, let me take you on the journey from external reality to internal worldview step-by-step by considering Figure 8A, which illustrates the first steps that phenomena go through in their progress to conscious awareness.

Tablemate: Although I appreciate the effort, if you were clearer in your arguments, I wouldn't need to suffer through all these diagrams.

Theorist: Possibly, but it would take a zillion more words to make my point. I started with a figure that progressed from external to internal, but the image was crowded, so I've broken it into three steps, starting with the sensory slice shown in Figure 8A, depicting the universe of all occurrences.

External to Internal
Sensory Slice

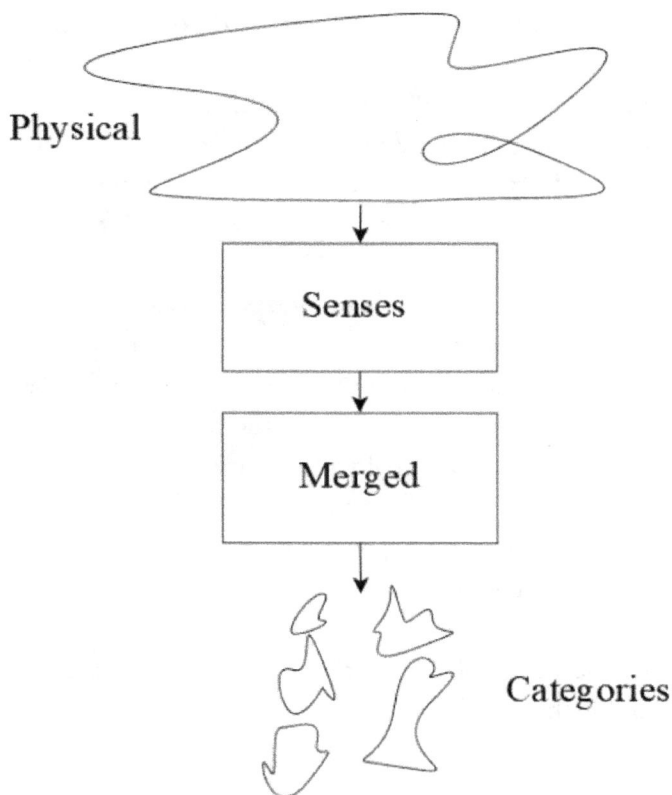

Figure 8A

IN THIS SIMPLIFIED graph, I want to emphasize that we experience but a sliver of the range of physical sensations, thereby missing not only simple events but also grand events. Obviously, we don't know what people say when we are not with them. On a simpler level, we do not experience light the same way as bees do. Clearly, for those that are color blind, the sensed reality would be vastly different compared to those that see color. On a grand scale, we miss out on all events except those occurring in our immediate vicinity.

Tablemate: Fiddle-faddle. Everybody knows that. Well, maybe some people don't know about the differences in light perception, but so what?

Theorist: Although we link events together as we experience them, we may not be aware of intermediate occurrences which are necessary for the full grasp of a situation, resulting in gaps in our understanding.

Tablemate: I beg your pardon. I don't go through life with broken connections.

Theorist: No, I'm sure you don't. We will discuss how we make those connections shortly, but for now I wish to convey that the conscious mind is not involved in the gathering of sensory data. Our sensory cells signal the spinal cord, then the brainstem, through the limbic system's thalamus, and finally to sensory enhancing capabilities of cortical lobes. The breadth of sensory events has been channeled into dedicated sensory processing regions. From the data that arrives, our brains extract color, objects, movement, distance, loudness, phonemes, odors, textures, and so on without conscious intention. These fragments are the sensed physical world. In Figure 8B, I will show you how we assemble the world into what I termed naïve reality.

External to Internal

To Naive Reality

Categories

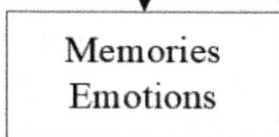

Memories
Emotions

Personal

Primitive

Figure 8B

THE ENHANCED SENSORY interpretation shown in Figure 8A
travels across neural pathways laid down during our learning phases.
These pathways trace categories we found useful in organizing and
reacting to the world, which includes the 3S Imperatives (Satiety, Sex,

and Safety). These categories provide immediate classification for situations we encounter.

Tablemate: Remind me, what exactly do categories include? Is my memory of the warmth of my grandmother's hug a category, or is that a combination of categories?

Theorist: It's both. Categories, as mentioned in the concept elevator discussion, pick up other aspects as they travel from sensory to association areas en route to the front lobes. As they move towards the cortical executive decision-making areas, categories—and concepts, as we may call the categories that are associated with theories and beliefs—become more abstract, less tied to the current environment and more yoked to one's experience.

The primitive mental worldview comprises categories and how those categories relate to one another, as determined by emotional history. Figure 8B shows three groups of categories in the primitive world. Clearly, three is not a maximum, but was chosen to simplify the graph and focus attention on the process. Since everyone's brain and personal experiences differ, the number of groups and their composition can easily differ.

The important takeaway is that, before we apply our learned knowledge, a rudimentary categorization already exists for the situation.

Tablemate: Come on. Clearer. I see my Power sandwich being made at the counter and I want to eat without worrying about how we apply our knowledge.

Theorist: Anything to oblige my favorite discussion partner. Before your lunch arrives, can you glance at the decision-making aspects of the internal worldview shown in Figure 8C?

External to Internal
World of Decision-Making

Naive
Reality

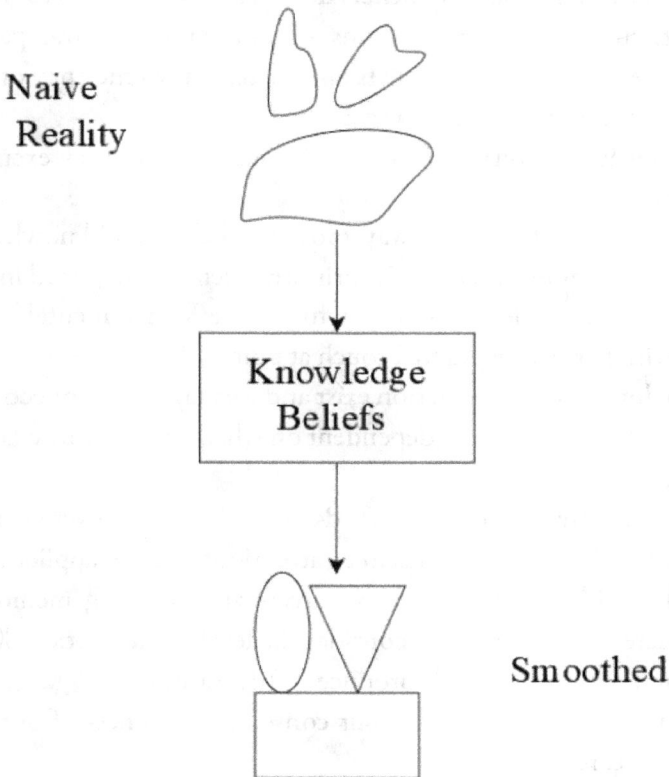

Knowledge
Beliefs

Smoothed

Figure 8C

OUR NEURONS PROCESS signals rapidly and automatically, forming neural pathways that transmit our primitive worldview augmented with knowledge. When we lack knowledge relating to categories, we typically rely on beliefs about relationships between categories. Beliefs are formed or learned in conversations or through media exchanges. Whether right or wrong, these beliefs fill the gap when knowledge is lacking. In fact, most beliefs are true in limited situations or under certain conditions, but the casual methods through which we learn and develop beliefs do not lend themselves to nuanced thought. Just as academic lessons strengthen the neural pathways among the facts conveyed, beliefs also increase the efficiency of neural pathways related to a particular concept.

Tablemate: I don't care for my beliefs being cast as excuses for knowledge.

Theorist: I feel the same way. However, beliefs and knowledge are subject to the same criticism. The primitive elements depicted in Figure 8C are shapeless and disjointed, while the effective mental world is shown with perfect shapes that touch at points. The goal is to show that isolated domains of information exist and we may see a connection that is narrow, constrained, and dependent on what we have learned or what we believe.

The effective worldview is developed and delivered to our prefrontal lobes, where conscious attention can be applied to the internal worldview. The executive areas with working memory can manipulate the worldview to consider the level of satisfaction different behavioral choices would produce. This satisfaction gauge is an amalgam of our emotions and our conscious awareness of our needs, goals, and fears.

Tablemate: Let me take these figures. I'd like to look them over before I run into you again. Ciao.

Day 9. Epitome of Thought

T heorist: Good afternoon, picture lady. Can you spare a few
minutes?

Tablemate: Again? I suppose I can take a break from actual work.
Entertain me with your flights of fancy.

Theorist: My principal goal is to understand how we think, and I
believe since we last met I've added a thing or two worth discussing.
So far, we haven't focused on the differences among preconscious,
unconscious, and conscious thoughts up to the higher-level thought
processes involved in conscious thinking, which is unique to humans.

Tablemate: Before you go on, what's the difference between
conscious and unconscious thought in your view?

Theorist: In the eons before humans came on the scene, organisms
learned to address their life's challenges such as finding food, attracting
partners, and avoiding predators. This was long before language and
words, and even longer before the cortical lobes expanded our mental
capacity. These behavioral responses are unconscious—or rather
preconscious—thought. I reserve the term 'unconscious' for those
thoughts arising in our emotions, within the limbic system, that are
difficult to put into words. Our brains have developed atop earlier
neural structures. Those activities originating in the deeper, older brain
activity are preconscious.

Tablemate: Emotions are different, aren't they? I sometimes do
things I wish I didn't. At other times I decide to do something but can't
explain why.

Theorist: Emotions are closer to consciousness, closer to the cortex
than our primitive responses to the 3S Imperatives. Emotions
originated 230 million years ago, before language processing in the

82

cortical lobe. The emotions lay deeper and are more archaic than cortical regions that control verbal abilities. After we make a choice driven by emotions, we can describe why we made such a decision—but only after the fact, by examining our internal worldview. Still, we are not describing the actual emotion, but the manner in which it shapes our worldview. Words are not directly linked to emotions, resulting in an uneasy fit.

Tablemate: Your answer raises many other questions. Here's a basic one. Why waste time on higher-order thinking when emotions can override the best logical conclusion?

Theorist: A tough but fair question. Do you remember our discussion about the 3S Imperatives (Satiety, Sex, and Safety) transmuted into emotions of needs, goals, and fears?

Tablemate: Stored in the amygdala, aren't they?

Theorist: Yes. Good memory. Emotions determine our level of satisfaction with events as we experience them. Our emotions shape the information the cortical hemispheres act on. Daniel Kahneman (2011) and Keith Stanovich (2009)[15] categorize higher thinking into two modes. System 1 is fast and always ready with a response to the current situation. System 2 is slow and requires effort, but may come to a superior conclusion. The fast mode is the pattern-feature-match-associative-substitution procedure which is nonverbal. The slow mode uses logic, acting on verbal categories.

Tablemate: Let me see if I understood you correctly. You believe that there are two unique thought processes in the brain. Are we merely a convenient battleground for the hemispheres?

Theorist: No, not at all. The brain provides a complementary result. Look at the layers of thoughts in Figure 9 depicting what I term Pyramid of Cognition akin to Maslow's hierarchy of needs.

Pyramid of Cognition

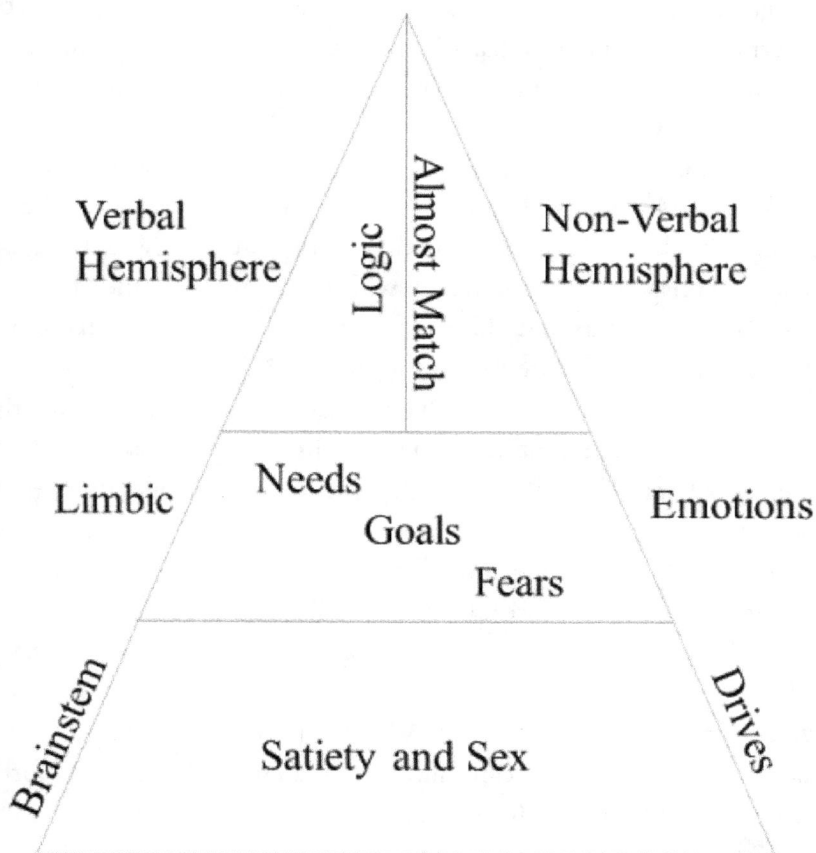

Figure 9

THEORIST: AT THE BASE, we have the oldest, non-conscious imperatives, driving our actions. Above that, the limbic system converts our experiences and consequences of our actions into emotions. Above that, the highest brain levels are divided into separate right and left hemispheric processing. Pattern processing occurs in the nonverbal hemisphere, without words. Data from both hemispheres are compared via the corpus callosum. Patterns (idiosyncratic categories derived from personal experiences) and words (common linguistic categories) rarely match exactly. That is why providing a verbal description of your personal insights can be difficult. You may even find your explanation drift to match the words used to describe what is nonverbal in your mind.

Tablemate: I can relate to that. That happened last week in the book club. Not to me, but to that artist that is sketching in the far table. His take on Marie Benedict's intention in *Carnegie's Maid* was so strong initially, yet he ended up with the tautology—the novel was good because it was good. But never mind that, let me ask, why should the hemispheres compare data as they process?

Theorist: The first reason is that one categorization may be superior to the other. One might align the situation more precisely with our needs, goals, and fears. If so, the other hemisphere changes its category to match as closely as it can to the better one. Second, the verbal hemisphere requires time and effort to arrive at its solution. When a response is essential before all logical possibilities can be examined, the nonverbal hemisphere provides a quick-and-dirty decision. However, this inductive path may overlook features lost in 'almost' matching that we later realize were important. When the situation is not pressing, the two executive regions can shuttle their latest results back and forth, each frontal lobe honing its outlook with the best fit of the two choices.

Tablemate: Both hemispheres include neurons and neural layers. So, the verbal hemisphere must proceed via almost matches as well. Isn't that contrary to the exact nature of logical true and false?

Theorist: How does the neural brain perform deductive logic? That's an excellent question which logic advocates should explain. It's a commonplace that we use logic, which is why I haven't focused on explaining logic's occurrence. My goal is to highlight the importance of feature matching, which leads to associations, induction, intuition, and creative aspects of human thought.

But realizing that logic arising from neural networks wasn't obvious, I have done a preliminary investigation. Using a Kohonen artificial neural network based on Caudill and Butler's algorithm (Understanding Neural Networks, 1992), I found that the basic logical operations—AND, OR, XOR, NOT, and IF–THEN—were learned[16] (formed a self-organized mapping) when exposed to sets of logical true–false values. Thereafter, the winning neuron in this artificial neural network found the correct logical operation in each case considered. Of course, that doesn't prove or demonstrate that the verbal executive area and working memory interact in this manner, but it shows that neural circuits can reliably apply logical assignments in a similar manner by which they categorize trees, letters, and emotions. Words and logical relationships become elements of thoughts as they pass through neural layers governed by almost gates.

Tablemate: You've lost me in your artificial neural network weeds. Let me ask you something else relating to words. Do you believe in the Sapir–Whorf Hypothesis, which holds that language dictates our reality?

Theorist: I do not believe in the hard version of the hypothesis. Our words don't define our worldview. We can always add qualifiers and adjectives to elaborate descriptions of our worldview. However, the soft hypothesis that language affects our worldview seems reasonable for verbal, logical thoughts. In *The Overflowing Brain: Information Overload and the Limits of Working Memory*, Klingberg (2008, p. 87) relates that constraints of working memory affect reading comprehension. It's barely a stretch to assume that requiring extra identifiers to achieve a fuller description consumes storage space designated for working memory and constrains mental capacity in other matters, including decision-making, reducing the scope for logical operations. One's language shapes the internal worldview to some extent, but it doesn't define it.

Tablemate: Wait a second. What about the concept elevator and neural cascade? As I recall, details were lost, causing increasing abstraction as categories rose towards the frontal lobes.

Theorist: That's right. Although there are some similarities, prefrontal cognition is different. With the neural cascade, loss of details occurs in deeper regions of the brain, prior to conscious activity. In the Sapir–Whorf Hypothesis (Philosophy of Linguistics, 2022), language constrains reality in the dominant hemisphere's executive area. The concept elevator describes the steps from preconscious to conscious thought in both hemispheres.

Tablemate: I have another concern. Are you saying we must choose between imperfect associations of the nonverbal hemisphere or assured deductions that might never complete? How could we ever decide anything?

Theorist: We choose to accept the risk of making a mistake based on false associations, or we guess at the assumptions needed to complete logical analysis.

This approach is taken in science, as hypotheses are formed and facts are analyzed with logic. The hypothesis is an inductive generalization arising in the non-dominant hemisphere, while deduction occurs in the verbal hemisphere. When they work together, our knowledge advances.

Day 10. Decision-Making

Theorist: Good morning. You seem deep in thought, should I find another table?

Tablemate: Oh, it's you. I'm arranging photos for the business phone book, or rather I'm trying to, but the process is driving me insane.

Theorist: Alphabetical order isn't good enough?

Tablemate: No. Nothing so simple. I have multiple pictures for each business. Do I use a front building shot for all businesses? Or only for some? Do I use daylight shots? But the neon-lit evening buildings pop better. Does the ABC Industries image show well next to a similar photo of AAA Corporation? To add to my dilemma, many new businesses only have a PO box to go with their telephone number and email address. They want a picture that captures their brand. I'm not sure what goforit.com offers, while party24by7.social has many possibilities, but I don't want to entice minors with the risqué image they provided.

Theorist: Is it a fair observation that your physical imperatives are met? Your mug is filled with chai. This is a comfortable table in the shade. We both have been vaccinated against COVID-19 and no one else is sitting too close. In addition, the decision doesn't have to be made this moment.

Tablemate: That's right.

Theorist: Thus, your emotional state is quiet. You're marshalling your conscious resources to a difficult problem.

Tablemate: My emotional state was quiet. I was calm and nothing was bothering me until you wandered over. I have a Monday deliverable.

Theorist: I can move if you wish.

Tablemate: No, it's too late. Let me tell you about the problem. First, I place the pictures this way, then I see a clash and try another way. But as that does not satisfy me, I go back to the first arrangement. My latest idea is to divide clients into subsets. Corporations get office interior photos. Individual storefronts have exterior shots. Government offices are still stymying me, which leaves cyber stores as a separate issue.

Theorist: You're clearly using both modes of thought, assigning verbal categories to intuitive patterns that make sense to you.

Tablemate: Lay it on me, scribbler. The almost gate will solve all my problems, right?

Theorist: Only in a general sense. Both word-based consideration and pattern matching occur in the upper reaches of the executive lobes. Both receive information filtered by almost gate operations that precede conscious awareness. Thus, the almost gate is not the differentiator.

Tablemate: Then what is?

Theorist: I have divided decision-making into four segments allowing them to be depicted by simple graphs, starting with Preconscious shown in Figure 10A.

Decision-Making

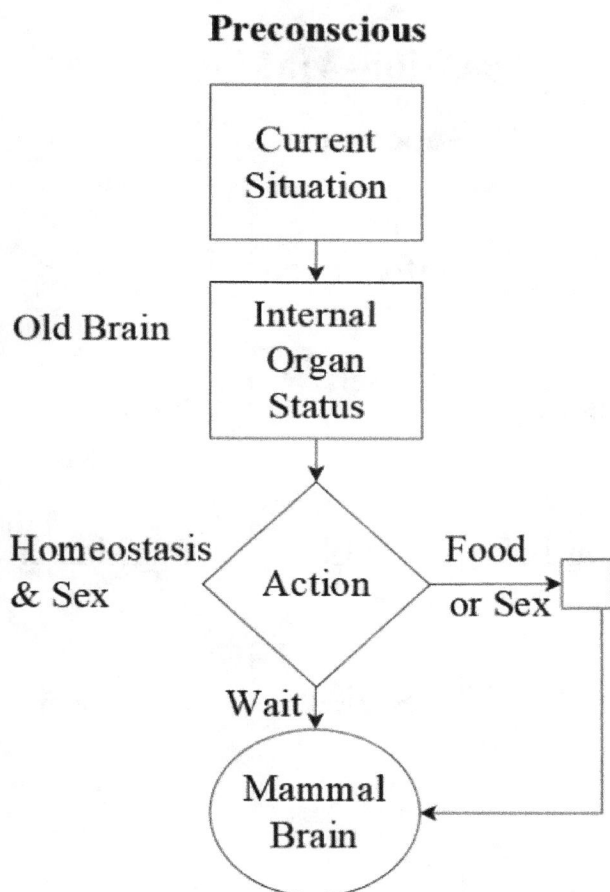

Preconscious

Old Brain

Current
Situation

Internal
Organ
Status

Homeostasis
& Sex

Action

Food
or Sex

Wait

Mammal
Brain

Figure 10A

YOUR PRECONSCIOUS MIND has already organized your
worldview into categories of interest with various levels of abstraction
and detail—corporations, government agencies, cyber storefronts, and
so on. If your 3S Imperatives weren't met, you would have to take care

of them, but as this is not the case, you can continue working on the business photobook problem. The preconscious worldview is handed over to the mammalian brain, as shown in Figure 10B.

Decision-Making

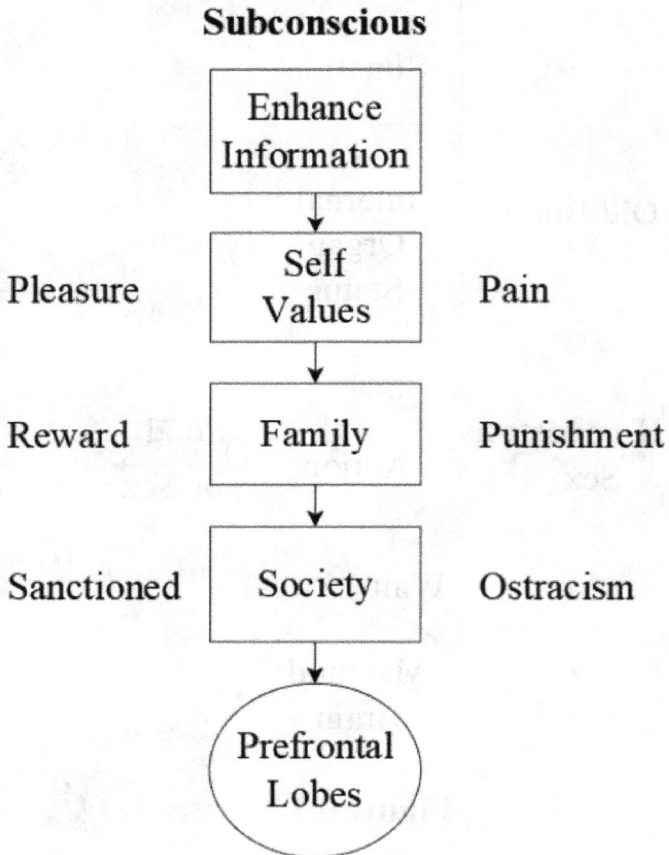

Subconscious

Pleasure   **Enhance Information**   Pain

Pleasure   **Self Values**   Pain

Reward   **Family**   Punishment

Sanctioned   **Society**   Ostracism

Prefrontal Lobes

Figure 10B

TABLEMATE: YES, YES. You went through the subconscious last time, but this looks somehow different.

Thesis: First, the large cortical lobes in humans focus most of their neurons on adding details and relationships between pieces of sensory information as the signals travel toward the prefrontal lobes. In the image, I highlighted the layers of categories and concepts a person learned on their journey to adulthood. The limbic structures add our emotional and moral judgments to the information.

The neural signals flow through pathways primed from our earliest pain–pleasure experiences, then through neural layer pathways shaped by family experiences weighed by reward–punishment feedback, and thence through pathways enriched by the results of our prior behavioral decisions interacting with society, earning approval, or ostracism.

As you look at the right-hand column in Figure 10B, you will notice that the source of our knowledge grows in breadth but shrinks in its certitude. We are positive about what we directly experienced, less sure about the knowledge our family imparts to us, and often at the mercy of the authorities that relate remote events to us.

The limbic system's amygdala uses our available knowledge to assign an emotional weight to the current situation. This occurs by shunting the situation through particular neural layers with synaptic gaps primed by experience.

Since your needs, goals, and fears are quiet, you feel calm and the mammalian brain delivers the situation without undue emotional weights to the conscious mind.

Tablemate: Why don't we always wait for logic to come up with the best solution?

Theorist: Maybe Figure 10C, Conscious Thought, will help you see the answer to that question.

Decision-Making

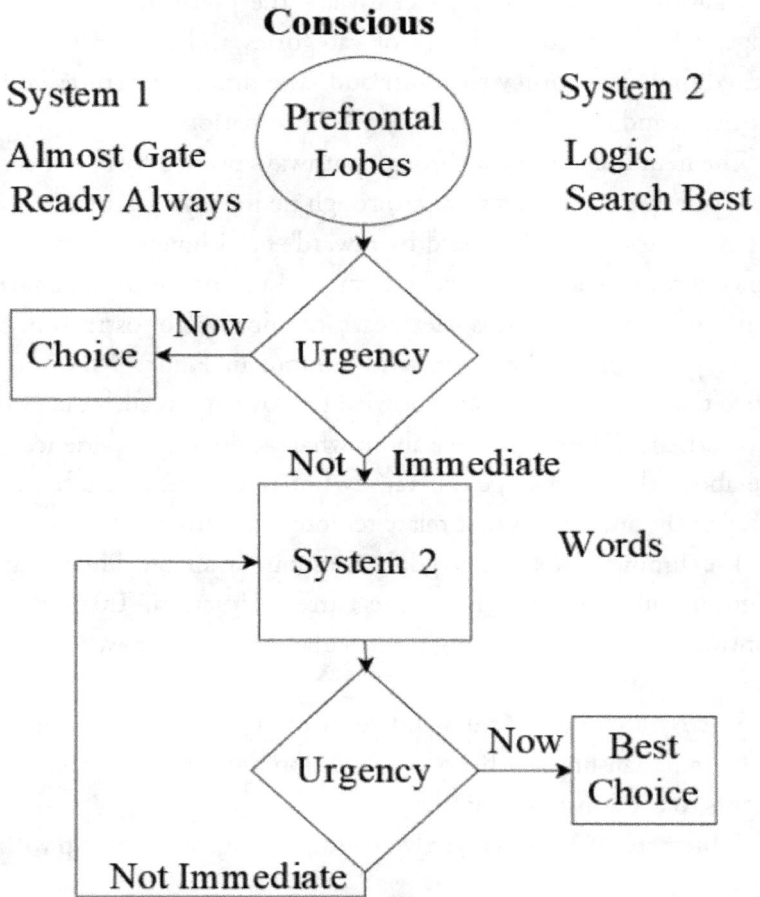

Conscious

System 1

Almost Gate
Ready Always

Prefrontal
Lobes

System 2

Logic
Search Best

Now

Choice ← Urgency

Not ↓ Immediate

System 2   Words

Urgency — Now → Best Choice

Not Immediate

Figure 10C

WHY DON'T WE ALSO WAIT for logic to come to the rescue and solve the problem? First, we often need to respond before logical thought has considered everything. Second, we may not have all the facts necessary to draw a conclusion, and important relationships between facts may be unknown. In either of the cases involving missing information, we use knowledge (Kahneman, 2011, pp. 79-88) or beliefs to fill in the gaps.

Tablemate: I never do any such thing.

Theorist: We all do. Consider this photobook. You don't have the knowledge of the relationship between the images and their acceptability to any of your clients. You separated the clients into arbitrary categories and guessed that their business perspectives would align.

Tablemate: Well, maybe when you put it like that I can see your point of view. Anyway, you do the same thing in these neural almost gate dialogs. You slip in unsubstantiated knowledge and beliefs. I'm sure you do.

Theorist: Guilty, I'm sure, though I'm pleased you can't name where off-hand. Notice the fast and slow thought modes, denoted as System 1 and System 2, at the top of Figure 10C. In the conscious mind, the hemispheric lobes have different specialties and responsibilities.

System 1 is always ready with a reaction. It operates on the prevailing internal worldview and appraises each situation through pattern matching and almost gates.

System 2 uses verbal categorization and neural layers honed for logical operations to search for the best decision. It starts with a copy of the internal worldview and current situation, then considers changes that could result from by certain actions, and then even further actions, in its quest for the best response to the current situation. That is a laborious, effortful process that demands time and attention.

The two systems work in tandem, communicate their results to the other. If time permits for all logical possibilities to be evaluated, the response that most closely matches our needs, goals, and fears will be chosen.

Tablemate: Wait. My 3S Imperatives are all met. My needs, goals, and fears are slaked.

Theorist: Not exactly and not forever. Though your primal needs are met, your emotional desires still exist. You may take this situation as an opportunity to consider how to further your professional success. If you work on this into the afternoon, hunger will arise. You might need a break to eat while your mind continues to work on it.

Tablemate: The mind working on it while not thinking on it. That sounds like subconscious.

Theorist: I'm glad you mentioned that, because System 1, the fast system, has its own lobe in the prefrontal cortex. Since it does not actively use words, but relies on patterns, System 1 doesn't communicate verbally. That makes it unconscious.

If System 1 comes upon an improved decision, System 2 sees it across the corpus callosum. We may mystically claim the subconscious delivered the solution, when, in actuality, the solution arose in the highly sophisticated, silent, pattern-matching prefrontal lobe.

The Response Urgency decision box is not static during conscious thought. As your hunger may divert you to eat and as Monday approaches, you are compelled to decide, even if uncertain of your choice.

In Figure 8C, after the decision box, System 2 is emphasized because it demands effort to continue, while System 1 does not. If an optimal solution is not at hand, we weigh the costs and benefits of the continuing effort. Everyone tires of the effort with an incomplete resolution at a different pace. The upshot is that, at some point, we may stop considering and settle for the latest of the System 1 and System 2 responses.

Tablemate: I'm halfway to there already.

Theorist: Once a person has chosen their response, whether by logic, pattern, urgency, or fatigue, that is not the end of decision-making, as feedback comes into play, as demonstrated in Figure 10D.

Decision-Making

Feedback

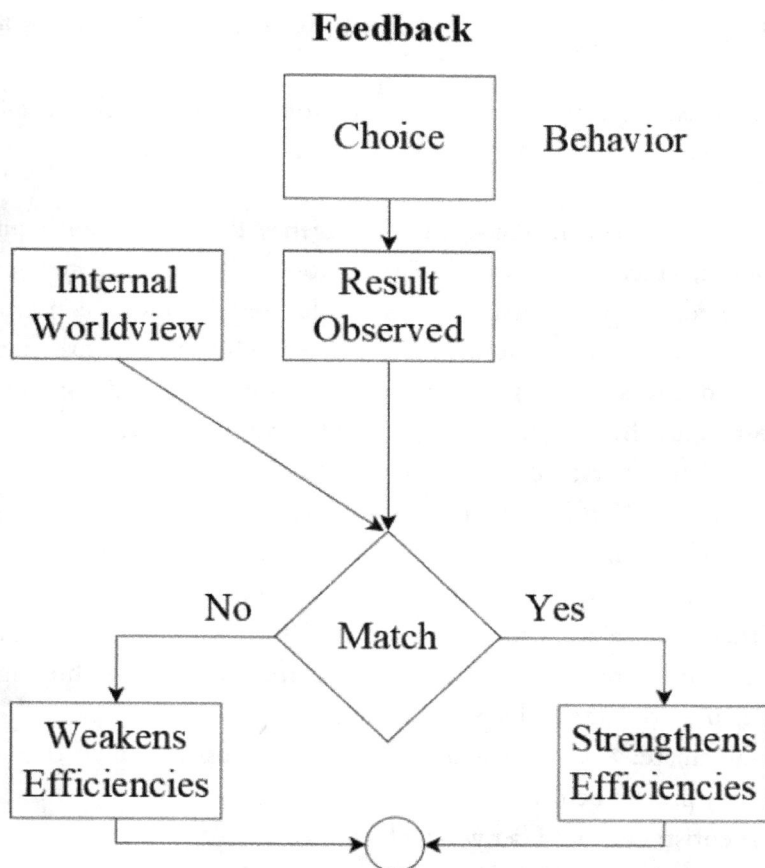

Figure 10D

WE AUTOMATICALLY NOTICE whether the expected outcome was achieved as a result of our conscious choice. The changed situation is evaluated, as we discussed earlier in relation to images depicting the transition from External to Internal, guiding us from Sensory Slice (preconscious) to Naïve Reality (emotional) to Decision-Making (conscious). The most noticeable feedback effort, if needed, will be adjustments to our worldview, reflecting changed synaptic efficiencies that align to the perceived reality changes.

Tablemate: Do you have to make things sound so dour? That reminds me of a question I want to ask. Why doesn't everyone think the same?

Theorist: Excellent question. Everyone does not think alike because we have different genetics and diverse experiences.

Tablemate: Please, not clichés.

Theorist: Let me try again. Our genes determine our neural threshold, which then obviously varies across individuals. As previously mentioned, people with higher thresholds require more significant feature matches to surmount almost gates, while those with lower thresholds are satisfied with few feature matches to surmount their almost gates. In addition, our particular experiences cause distinct patterns in our personal categories.

Tablemate: Do I detect the old Nature–Nurture debate? I'm not sure there's anything you have said that would convince anyone in the coffee shop to come to a new understanding of that argument.

Theorist: I neither believe nor intend for my neural almost gate explanations to put all issues to rest, but perhaps a recap will highlight an idea or two that may be worthy of remembering.

Tablemate: Are you concluding? That's a positive omen that I'll finish my project before Monday.

Theorist: Yes, and I bet you will finish it too.

The primary insight from all our discussions so far is that the almost gate arises from the neural threshold delivering the same signal regardless of the inputs that triggered it. In layers, the almost gate results in the self-organization of winning neurons into memory maps of categories and concepts.

As 99.9% of the brain's neurons (Spitzer, 1999) communicate with each other rather than the outside world, appreciating that a vast number of almost gates lead to category abstraction and loss of detail is crucial. The concept elevator is a useful metaphor for the amalgamation of memories and emotions into abstract categories to make their way to decision-making areas. The divergence of individual reactions to the present is traced to the loss of sensory fidelity with the growth in importance of memory and knowledge.

Our 3S Imperatives (Satiety, Sex, and Safety) arose long before conscious thought, allowing primitive organisms to deal with their challenges. Even the emotion-handling sections of the brain are ancient, for they emerged more than 200 million years ago with the mammals. Back then, our ancestors learned to balance their needs, goals, and fears long before they acquired language. These emotions are preconscious and can only be described by their effects as we look at our choices. Emotions flavor the calculations of our conscious mind with urgency.

Tablemate: I do like the concept elevator.

Theorist: That's one thing. Yay. Better than these dialogs adding up to nothing.

A last bit. The verbal hemisphere truncates the world into language categories. It's logical and can imagine alternate possibilities. But it demands time and mental effort to do its work. The nonverbal hemisphere captures the experienced world in personal, idiosyncratic patterns. It's associative and deals with the here-and-now. Its goal is a ready response for every situation that confronts us, although it may miss distinctions later revealed to be important. Both modes of higher thought are shaped by our emotional states. The two hemispheres communicate with each other across the corpus callosum as they move their results toward the executive centers of the prefrontal cortex. This allows the most fruitful description (whether verbal or idiosyncratic pattern) to cross to the other hemisphere and become the base for further processing.

98

Tablemate: I never used to think about how we think, and I don't know how I will use these ideas. In the meantime, how about we share a biscotti order, scribbler?

Glossary

Abstraction. Loss of detail. Abstraction occurs at every almost gate. Every neuron in the human brain has an almost gate, as does every interconnected neural layer.

Almost gate. There are two types of almost gates, at a neuron level and in a neural layer.

1. The first almost gate occurs when the neuron's threshold is exceeded. Since the same signal is sent, regardless of minor differences in the inputs, the almost match is performed.
2. The second almost gate occurs in neural layers that are connected to the same input cells. A winning neuron is selected based on the closest match to the features of the layer's input set. The layer thus acts as an almost gate.

Association. The substitution of one pattern, category, or concept for another in thoughts. Association is generated when two thought entities surmount the same almost gate, yet a decision is not yet required. The executive areas may pursue related memories for additional alternatives. The association may evoke fruitful ideas from details that didn't exist in the original pattern, category, or concept. A metaphor is a verbal example of an association.

Belief. Mental acceptance of and conviction in the truth, actuality, or validity of something. Beliefs often arise from three main sources. Early experiences yield self-image, religion imparts morals, and science insights into nature.

Category. A linguistic group, verbal description of a pattern.

Concept. Categories that have meaning added by experienced relationships, learned knowledge, or beliefs.

Concept elevator. Metaphor for the routine changes that sensations pass through in their movement, through neural layers, from raw sensations to final understanding in the prefrontal lobe.

Critical learning period. If an implicit skill is not learned before the end of the critical period, the individual will never master it despite further attempts. This may be noticed in children who have been deprived of sensory inputs or social interactions.

Explicit learning. Learning that occurs primarily through verbal instruction. Schooling is a classic example. All conversations have the potential to be instructive, as lessons can be learned from parents, family members, friends, news, and gossip. Explicit learning can continue throughout a person's life, unlike implicit learning which occurs during a particular stage in a person's lifetime.

Implicit learning and skills. Basic skills, such as vision, hearing, bodily movement, language, and social skills that occurs through SOMs (Self-Organizing Maps). Much implicit learning starts before language is mastered. Implicit learning is characterized by critical learning periods.

Induction. The generalization of specific facts to universal statements. When a neural layer's winning neuron signals, its inputs are inducted into the same universal statement (or category) as all the other input sets that select that winning neuron. Receivers of that axon signal have generalized a range of input sets to one signal for subsequent brain processing.

Internal worldview. The world construction on which we base our decisions. It is unique because of a person's genetics and distinct experiential history.

Knowledge. Familiarity, awareness, or understanding gained through experience or study.

Neural cascade. As concepts proceed up the concept elevator, there is a gradual loss of signal fidelity because of passing through many almost gates.

Pattern. Set of features sufficiently alike that the same winning neuron of a neural layer is selected.

Pattern matching (aka feature matching). Two patterns match when an almost gate fires on the receipt of either pattern. The patterns may differ in unimportant details, but the preponderance of features match.

Self-Organizing Map (SOM). A layer of neurons that arranges inputs into patterns or categories based on feature similarity. SOMs support implicit learning that starts prior to explicit training that is word-based and instructor-led.

Situation. Our current reality. The situation is a perturbation to the internal worldview, which may require action depending on needs, goals, and fears.

System 1 and **System 2.** Handy terms used to describe separate capabilities of the human brain which occur in parallel. Although dispute exists over the exact domain of the two systems, System 1 is considered to be fast, inductive, non-verbal, automatic, and always ready with an answer, while System 2 is slow, deductive, verbal, requires effort, and sometimes cannot reach a conclusion.

3S Imperatives. Satiety, Sex, and Safety are drives built into our genetics long before consciousness. The limbic system synthesizes the 3S imperatives into emotions, with axes of needs, desires, and fears.

Winning neuron. In a neural layer, the neuron most activated by a set of inputs. The winning neuron can be triggered by multiple input sets, delivering its axon signal to subsequent neural layers.

Working memory. Area of short-term memory in the prefrontal lobes used to manipulate long-term memories that are germane to a current situation. Facts, beliefs, and knowledge in working memory can be considered in numerous arrangements, allowing new conclusions to be reached, which can be saved to long-term memory. This feature supports explicit learning into adulthood, as working memory retains its ability to adjust to new categorizations and concepts.

References

Aristotle. (1952). *Great Books, De Poetica* (Vols. 9, Aristotle II). (I. Bywater, Trans.) Chicago, IL, USA: The University of Chicago.

Baddeley, A. (2004). *Your Memory: A User's Guide.* Buffalo, NY, USA: Firefly Books.

Bermundez, J. L. (2003). *Thinking without Words.* New York, NY: Oxford University Press.

Carter, R. (1998). *Mapping the Mind.* Berkeley: University of Califormia Press.

Carter, R. (2009). *The Human Brain Book: An Illustrated Guide to its Structure, Function, and Disorders.* London: DK.

Caudill, M., & Butler, C. (1992). *Understanding Neural Networks* (Vol. 2). MIT Press.

Dweck, C. (2007). *Mindset: The New Psychology of Success.* New York: Ballantine Books.

Encyclopedia Britannica. (2022, 02 17). *Stages of Moral Development.* Retrieved from Encyclopedia Brittannica: https://www.britannica.com/science/Lawrence-Kohlbergs-stages-of-moral-development

Famous Scientists biography. (2022, 02 16). Retrieved from Famous Scientists: https://www.famousscientists.org/wilder-penfield/

Fields, R. D. (2009). *The Other Brain: From Dementia to Schizophrenia*. New York: Simon & Schuster.

Garlick, D. (2010). *Intelligence and the Brain: Solving the Mystery of Why People Differ in IQ and How a Child Can Be a Genius*. Burbank: Aesop Press.

Hadamard, J. (1954). *The Psychology of Invention in the Mathematical Field*. New York: Dover Publications.

Hebb, D. (1949). *The Organization of Behavior: A Neuropsychological Theory*. John Wiley and Sons.

Hebbian Learning. (2022, 02 21). Retrieved from The Decision Lab: https://thedecisionlab.com/reference-guide/neuroscience/hebbian-learning/

Heuer, R. J. (1999). *Psychology of Intelligence Analysis*. Reston, Virginia: Pherson Associates, LLC.

Hofstadter, D., & Sander, E. (2013). *Essences and Surfaces: Analogy as the Fuel and Fire of Thinking*. New York: Basic Books, Perseus Books Group.

Kahneman, D. (2011). *Thinking Fast and Slow*. Farrar, Straus and Giroux.

Kalat, J. W. (2004). *Biological Psychology* (8th ed.). Belmont, California: Thomson Wadsworrth.

Klawans, H. M. (1988). *Toscanini's Fumble and Other Tales of Clinical Neurology*. Chicago, IL: Contemporary Books.

Klawans, H. M. (2000). *Strange Behavior: Tales of Evolutionary Neurology.* New York: W.W. Norton & Company.

Klawans, H. M. (2000). *Strange Behavior: Tales of Evolutionary Neurology.* New York: W.W. Norton & Company.

Klingberg, T. (2008). *The Overflowing Brain: Information Overload and the Limits of Working Memory.* Oxford University Press.

Koestler, A. (1965). *The Art of Creation.* New York: The Macmillian Company.

Kohonen, T. (2014). *MATLAB Implementations and Applications of the Self-Organizing Map.* Aalto: School of Science, Aalto University.

Philosophy of Linguistics. (2022, 20 18). Retrieved from Stanford Encyclopedia of Philosophy: https://plato.stanford.edu/entries/linguistics/#Who

Pinker, S. (2007). *The Language Instinct.* New York: Harper Collins Perennial Modern Classics.

Schmidt, A. (2000, 10 04). *Biological Neural Networks.* Retrieved from Technology for Pervasive Computing: https://www.teco.edu/~albrecht/neuro/html/node7.html

Spitzer, M. (1999). *The Mind Within the Net. Models of Learning, Thinking, and Acting.* Cambridge, MA: The MIT Press.

Stanovich, K. (2009). *What Intelligence Tests Miss: The Psychology of Rational Thought*. New Haven: Yale Univeristy Press.

Sternberg, R. J. (Ed.). (1999). *Handbook of Creativity*. Cambridge University Press.

Sternberg, R. J., & Sternberg, K. (2009). *Cognitive Psychology* (6th ed.). Belmont, California: Wadsworth Cengage Learning.

Stull, R. B. (2000). *Meteorology for Scientists and Engineers*. Belmont, CA: Brooks/Cole.

The Reptilian Brain. (2022, 02 21). Retrieved from Current Biology: https://www.cell.com/current-biology/fulltext/ S0960-9822(15)00218-3

Zeman, A. (2008). *A Portrait of the Brain*. New Haven: Yale University Press.

Image Citations

1. Front cover. Neuron system complex model. GarryKillian. iStock Illustration ID 1277789824
2. Figure 1. Structure of a Typical Neuron. Public Domain, https://commons.wikimedia.org/w/index.php?curid=254226
3. Figure 6. Pheasant Tracks. Walter Baxter cc-by-sa/2.0 www.geograph.org.uk/photo/1058324. Website.
4. In Endnote 5. Penfield Homunculus from Wikimedia Commons. https://commons.wikimedia.org/wiki/Template:BA312_-_Primary_Somatosensory_Cortex_-_with_homunculus
5. In Endnote 6. Overview of brain structure from Wikimedia Commons, courtesy of Bruce Blaus - Own work, CC BY 3.0. https://commons.wikimedia.org/w/index.php?curid=31118589
6. All other figures were developed by the author.

Endnotes

[1] I discuss Facts Defined. Distinguished from Explanations, Evaluations, and Predictions[1] at https://www.mentalconstruction.com/knowledge-defined-classified/.

[2] Hebb's Law or Rule, proposed by neuropsychologist Donald Hebb, states that synaptic connections are strengthened when two or more neurons are activated contiguously in time and space. When the firing of the presynaptic cell is associated with the activity of the post-synaptic cell, structural changes take place that favor the appearance of assemblies or neuronal networks. The Virtual Psych Center website has fine coverage of Hebb's Law's[2] influence in neuropsychology at https://virtualpsychcentre.com/hebbs-law-the-neuropsychological-basis-of-learning/.

[3] Wikimedia Commons shares this excellent image of a synaptic gap.

[4] In the How similar is similar enough?[3] post, I develop an inequality that needs to be solved. https://www.mentalconstruction.com/similar-enough/

$$\sum_{k=1}^{7000} (firing_k efficiency_k) - \sum_{m=1}^{3000} (firing_m efficiency_m) \geq Almost\ Gate\ threshold$$

[5] Wikimedia Commons image[4] that displays Penfield's discovery of dedicated cortical areas to receive, categorize, and respond to tactile data, known as the Penfield

---

1. https://www.mentalconstruction.com/knowledge-defined-classified/

2. https://virtualpsychcentre.com/hebbs-law-the-neuropsychological-basis-of-learning/

3. https://www.mentalconstruction.com/similar-enough/

4. https://commons.wikimedia.org/wiki/

Template:BA312_-_Primary_Somatosensory_Cortex_-_with_homunculus

homunculus.

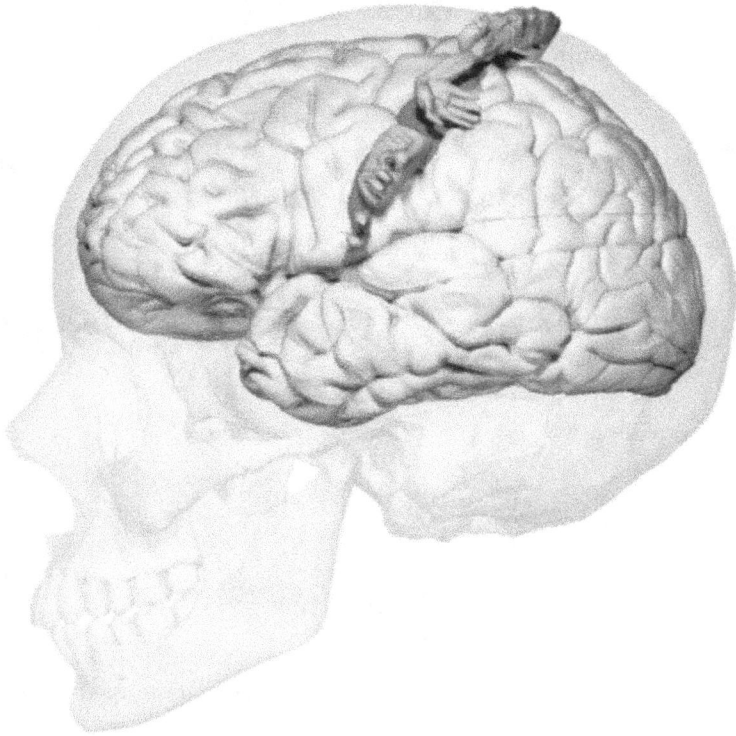

[6] Brain structure by function
Occipital lobe handles vision
Temporal lobe handles hearing
Parietal postcentral gyrus handles touch
Frontal precentral gyrus handles movement

Lateral View of the Brain

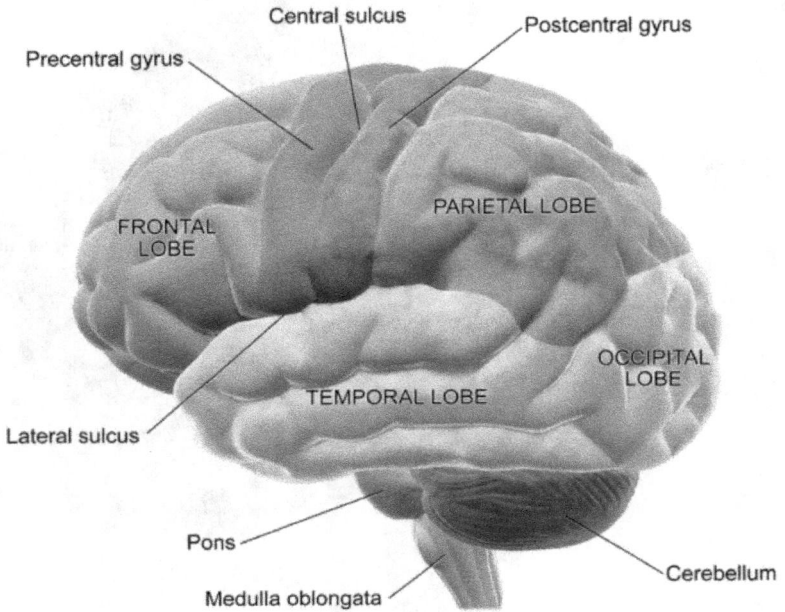

Central sulcus

Postcentral gyrus

Precentral gyrus

PARIETAL LOBE

FRONTAL
LOBE

OCCIPITAL
LOBE

TEMPORAL LOBE

Lateral sulcus

Pons

Cerebellum

Medulla oblongata

[7] BrainPaths website provides additional background[5] on the mechanism of Hebb's Law.

[8] Implicit learning and implicit memory are forms of learning and memory that occur without the person's awareness and depend upon different brain systems from those underlying consciously controlled (or 'explicit') learning and memory.

[9] Neural layers or modules are subdivisions of the Brodmann areas created by dividing the cerebral cortex into 52 distinct segments based on their cytoarchitecture as well as various functions. https://en.wikipedia.org/wiki/Cerebral_cortex#Cortical_areas

[10] The letter E in various fonts and scripts.

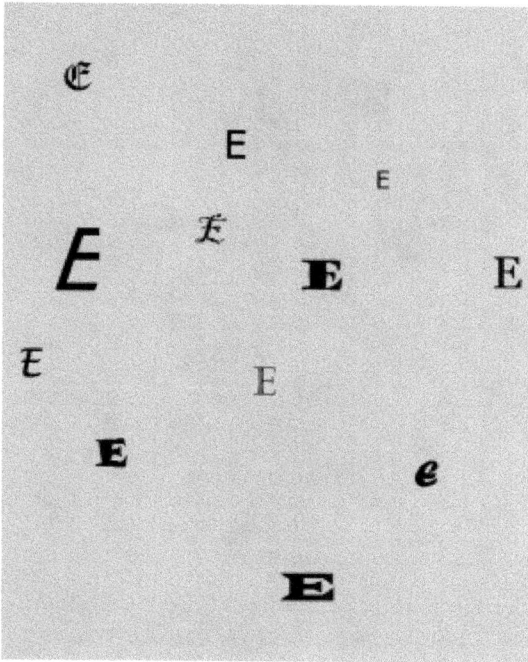

[11] Only in the prefrontal executive areas does working memory allow us to consider alterations to future situations that depend on actions that we might take now, not just one step ahead, but further steps contingent on the alterations caused by the possible prior behaviors.

Consciousness is the ability to distinguish between what is happening now and what might be occurring later. As we all know, sometimes the allure of what might be becomes more important than the reality that is.

[12] Knowledge and beliefs are similar but not the same. Knowledge is an organized collection of facts and their interrelations. If a fact contradicts the knowledge set, something must change. Beliefs are also an organized collection of facts and their interrelations. However, if a fact contradicts a set of beliefs, it can be set aside, declared the result of a supernatural intervention, leaving the belief set intact.

[13] A valid argument exists if all premises that are true lead to a conclusion that is true.

A sound argument is an argument that is valid, and all of its premises are true.

[14] Kohlberg's Stages of Moral Development graphically displayed. Although alternatives to Kohlberg's staging exist, his insight of incremental development is fruitful.

Kohlberg's Theory

Level/Stage	Age Range	Description
I: Obedience/Punishment	Infancy	No difference between doing the right thing and avoiding punishment
I: Self-Interest	Pre-school	Interest shifts to rewards rather than punishment – effort is made to secure greatest benefit for oneself
II: Conformity and Interpersonal Accord	School-age	The "good boy/girl" level. Effort is made to secure approval and maintain friendly relations with others
II: Authority and Social Order	School-age	Orientation toward fixed rules. The purpose of morality is maintaining the social order. Interpersonal accord is expanded to include the entire society
III: Social Contract	Teens	Mutual benefit, reciprocity. Morally right and legally right are not always the same. Utilitarian rules that make life better for everyone
III: Universal Principles	Adulthood	Morality is based on principles that transcend mutual benefit.

The Psychology Notes Headquarters - http://www.PsychologyNotesHQ.com

[15] In his text, Stanovich refers to the shortcomings of IQ tests to elaborate on his theme, captured in the book's subtitle, *The Psychology of Rational Thought*.

[16] A brief summary[6] of the experiment is available at https://www.mentalconstruction.com/mental-construction/cognition/reasoning/deduction-neural-brain/.

---

6.   https://www.mentalconstruction.com/mental-construction/cognition/reasoning/deduction-neural-brain/

About the Author

For thirty years, Robert Hamill has studied the advances in neural science and wondered at the lack of a theory explaining intuition, inspiration, and creativity.

Every day since his first presentation that neural networks categorized without training and without logic, he has been growing the theory of the almost gate and how it pervades the shape of our thoughts.

This book includes his theory of non-logical and creative idea formation. Parts of these ideas have appeared in Mensa publications.

Read more at https://www.rhamill.com.

www.ingramcontent.com/pod-product-compliance
Lightning Source LLC
Chambersburg PA
CBHW060506280326
41933CB00014B/2880